THIRD GENERATION
WIRELESS INFORMATION
NETWORKS

THE KLUWER INTERNATIONAL SERIES
IN ENGINEERING AND COMPUTER SCIENCE

COMMUNICATIONS AND INFORMATION THEORY

Consulting Editor:
Robert Gallager

Other books in the series:

Digital Communication *Edward A. Lee, David G. Messerschmitt*
ISBN: 0-89838-274-2

An Introduction to Cryptology *Henk C.A. van Tilborg*
ISBN: 0-89838-271-8

Finite Fields for Computer Scientists and Engineers *Robert J. McEliece*
ISBN: 0-89838-191-6

An Introduction to Error Correcting Codes With Applications
Scott A. Vanstone and Paul C. van Oorschot
ISBN: 0-7923-9017-2

Source Coding Theory *Robert M. Gray*
ISBN: 0-7923-9048-2

Switching and Traffic Theory for Integrated Broadband Networks
Joseph Y. Hui
ISBN: 0-7923-9061-X

Adaptive Data Compression *Ross N. Williams*
ISBN: 0-7923-9085-7

Advances in Speech Coding *ed: B.S. Atal, V. Cuperman, A. Gersho*
ISBN: 0-7923-9091-1

THIRD GENERATION WIRELESS INFORMATION NETWORKS

Edited by

Sanjiv Nanda
AT&T Bell Laboratories

David J. Goodman
Rutgers University

KLUWER ACADEMIC PUBLISHERS
Boston/London/Dordrecht

Distributors for North America:
Kluwer Academic Publishers
101 Philip Drive
Assinippi Park
Norwell, Massachusetts 02061 USA

Distributors for all other countries:
Kluwer Academic Publishers Group
Distribution Centre
Post Office Box 322
3300 AH Dordrecht, THE NETHERLANDS

Library of Congress Cataloging-in-Publication Data

Third generation wireless information networks / edited by David J.
   Goodman, Sanjiv Nanda.
          p.   cm. -- (The Kluwer international series in engineering and
   computer science. Communications and information theory)
     Includes bibliographical references and index.
     ISBN 0-7923-9218-3 (Alk. paper)
     1. Telecommunication systems. I. Goodman, David J., 1939-   .
   II. Nanda, Sanjiv, 1961-   . III. Series.
   TK5103.T48   1991
     621.382.--dc20   91-32551
                                                                    CIP

Copyright © 1992 by Kluwer Academic Publishers

Printed on acid-free paper.

Printed in the United States of America

CONTENTS

Preface

Rutgers University launched WINLAB in 1989, just as the communications industry, the Federal government, and the financial community in the United States, were waking up to the growing public appetite for wireless communications and to the shortage of technology to feed it. The secret was already out in Europe, where no fewer than three new cordless and cellular systems were progressing from drawing board to laboratory to factory to consumers. In July 1989, the FCC held a well-attended tutorial that turned into a debate over whether second generation British or Swedish technology held the key to mass-market personal communications. Many in the audience wondered whether United States technology was out of the picture.

Technology uncertainties are more acute in wireless communications than in any other information service. For example multi-gigabit optical fiber communications have followed an orderly progression from basic science leading to technology, which in turn stimulated standards, and then commercial products. Eventually applications will be found and industry and society at large will reap the benefits. By contrast, the applications of wireless communications are apparent to an eager public. A large market exists but is held in check by a shortage of capacity. The demand has led the cellular industry to formulate standards for advanced systems before the technology is in place to implement them. Everyone holds their breath waiting to observe performance of the first products. Gaps in basic science add to the uncertainty and forestall the resolution of technological debates.

While this state of affairs produces risk and excitement in the short term, it also directs attention to the distant future, when public requirements will be much more exacting than they are today. Where will the technology come from to meet the needs of the next century? Who will create it and who will exploit it commercially? To provide a focus for addressing these questions in the United States, WINLAB, early in 1989 set up shop as an educational resource, a sounding board for new ideas, and an incubator of third generation technologies. The vision of the third generation is a compatible set of linked networks that allow people and machines in all locations to exchange all kinds of information.

In June 1989, the first WINLAB Workshop on Third Generation Wireless Information Networks reflected the preliminary status of long term research. A large proportion of the contributions covered emerging, second generation technologies and only five presented University research. In the sixteen months between the two WINLAB Workshops, as everyone scrambled to understand the technological and business choices for the early nineties, scores of workshops, seminars and tutorials were held throughout the country. Meanwhile, the WINLAB Workshop remains the only one focused on the long-term future of wireless communications.

At the October 1990 Workshop, the number of University contributions doubled, and the presentations and discussions addressed the technological uncertainties about third generation wireless information networks (WIN's). This book contains written versions of nineteen of the twenty-six presentations at the 1990 WINLAB Workshop. While the subject is the technology of WIN's, this technology, more than many others, is the product of an intricate matrix of business, regulatory, scientific, and sociological factors. The early chapters deal with the regulatory environment in which technology is created. Laurence Movshin, an attorney at Thelen, Marrin, Johnson and Bridges, describes the complexity of the issues that face United States government regulators and the diverse approaches at their disposal in charting the future of wireless services. Shila Heeralal, of Bell Cellular, steps back from the current scene and presents the case for international standards in the light of previous experience, social needs, business considerations, and stimuli for technology creation. Addressing current activities, Michael Callender of MPR Teltech, and Chairman of an influential Task Group within the International Telegraphic Union, describes deliberations on spectrum requirements to be presented to the World Administrative Radio Conference in 1992. The examination of standards concludes with Chapter 4 by Alistair Munro of the University of Bristol, which studies the application of the Open Systems Interconnection protocol to wireless information networks.

Chapters 5 and 6 deal with fundamental natural obstacles to the progress of wireless communications, shortage of bandwidth and shortage of power in wireless terminals. Harry Bims and John Cioffi of Stanford University address both issues with a reduced complexity realization of a bandwidth efficient coded modulation scheme. They propose a precoding scheme for trellis-coded continuous phase modulation, that permits Viterbi decoding with a considerably reduced state space. With "multimedia" a buzzword for future information services, people on the move will carry terminals containing sophisticated communication subsystems and far more processing capability than today's laptops. To make this possible with small batteries, a new approach to integrated circuits and algorithms will be essential. In Chapter 6, Anantha Chandrakasan, Samuel Sheng and Robert Brodersen from the University of California at Berkeley show that high speed operation and low power dissipation are contradictory and that parallel or pipelined processing will be essential in advanced terminals.

Approximately one-third of this book addresses the related topics of resource allocation and handoffs, two issues that are of utmost importance for future wireless systems. Although these issues elicited the largest response in the Workshop call for papers, no consensus emerges from the diverse approaches. To us this indicates that the issues are wide open, the problems are still being defined, and that the work presented is preliminary. The chapter by Sanjiv Nanda and David Goodman from WINLAB at Rutgers University describes allocation of TDMA carriers to base stations in response to temporal and spatial variations in traffic. To evaluate dynamic channel allocation techniques, this chapter offers a model for non-uniform traffic. Chapter 8 by Srikanta Kumar of Northwestern University, and Hwan Chung and Mohan Lakshminarayan of BNR, provides stochastic bounds to compare fixed channel allocation schemes with channel borrrowing. H-C Tan, Mustafa Gurcan and Zenon Ioannou of Imperial College discuss a TDMA slot allocation scheme for radio local area networks (R-LAN's) carrying packetized data. The push for R-LAN's is also discussed later in the book, in Chapter 17 by Ray Simpson. In Chapter 10, Izumi Horikawa, Masaaki Hirono and Kazusinghe Tanaka of NTT discuss a possible architecture for a microcell TDMA system with mobile transceivers that are capable of switching rapidly between different TDMA carriers.

In Chapter 11, Stephen Rappaport of the State University of New York at Stony Brook presents very general, computationally tractable analytical models for handoffs. This is a fresh approach to a problem that has previously been studied by means of simulations or measurements. In Chapter 12, Bjorn Gudmundson and Olle Grimlund of Ericsson Radio Systems also consider the handoff issue. Analyzing propagation measurements, they conclude that the handoff problem in microcells is a thorny one. They fit a simple model and perform simulations to study the effects of handoff thresholds and hysteresis. J.-F. Wagen of GTE Laboratories develops a model of radio wave propagation in microcells using the spectral incremental propagation procedure to calculate the electromagnetic field through and around obstacles.

Multiple access protocols that assign radio channels to spatially distributed wireless terminals must match the requirements of specific applications. David Chan of Mobile Data International analyzes the performance of a protocol designed specifically for a mobile packet data system. Chapter 15 by Bach Ngoc Bui, Philippe Godlewski and Sami Tabbane of Telecom Paris, discusses a protocol for direct mobile-to-mobile communications. In this case, the application is inter-vehicle communication for collision avoidance. Zenon Ioannou, Mustafa Gurcan and H-C Tan of Imperial College study a multiple access technique for data packets in a radio local area network (R-LAN), with access coordinated by a central base station.

Departing from the preliminary, tentative tone of much of the book, the final three chapters address current and emerging technologies. In Chapter 17, Ray Simpson of O'Neill Communications considers the network requirements of a small office and discusses how they can be fulfilled by low-cost, unlicensed spread spectrum communications. At this Workshop as elsewhere, the subject of cellular applications of spread spectrum provoked lively debate. While military applications have been around for decades, commercial implementations are fairly recent. Military applications are motivated by immunity to jamming, with the terminal cost and weight secondary considerations. Recent proposals for cellular applications have raised questions about cost, multiple access capacity, and the performance of systems in practical operating environments. The chapter by Allen Salmasi of Qualcomm describes his company's highly publicized code division multiple access (CDMA) cellular system. Salmasi shows that CDMA naturally takes advantage of antenna sectorization and speech activity detection. These two technologies play a major role in calculations that show CDMA capacity to be higher than that of TDMA. To achieve this capacity, however, dynamic power control is essential. This is the subject of the final chapter, in which Jack Holtzman of WINLAB mathematically relates power control accuracy to CDMA capacity.

While the scope of this book suggests the diversity of topics that require the attention of designers of third generation wireless networks, there are a number of research issues that receive insufficient coverage here. Communication theory for the wireless environment is a very rich subject that calls for continuing research. New schemes and enhancements of old ones are proposed in every conference. We look to the research community to determine the performance and efficiency limits of practical wireless systems. The inconclusive nature of the major portion of this book that is dedicated to handoffs and dynamic channel allocation stems from the large number of complex factors that affect performance. We expect in future Workshops to learn of increasingly sophisticated work that takes into account the combined influence of user mobility (road and pedestrian traffic), communication patterns (teletraffic), radio propagation, cell engineering, and algorithm design.

We expect the different multiple access schemes that today attract much of the attention of the wireless network community, to sort themselves out as performance comparisons come into focus. Meanwhile, as new information services emerge and signal processing technologies evolve, new multiple access technologies will come on the scene and stimulate their own debates. No doubt we will hear about them at future WINLAB Workshops.

In formal and informal discussions, Workshop delegates pointed to the absence of contributions on network architectures and signalling systems designed to provide high quality information services to mobile terminals communicating through unreliable radio channels. Many people believe that future networks will prosper or fail on the strength of the network infrastructure. Software intensive switches and infrastructure protocols to track user mobility will be more costly to establish and more difficult to change than the radio systems that absorb much of our attention today. We hope that these topics will take the spotlight in future Workshops.

We are privileged to have the opportunity to bring together, in one volume, a rich variety of studies of wireless information networks. Full credit is due to the authors who offer us their view of the future. We compliment them, not only on their research but on their achievements of clear exposition and attractive visual presentation. We are also pleased to acknowledge the dedication to this project of Stephanie Faulkner and Tom Stone of Kluwer Academic Publishers.

May 1991   Sanjiv Nanda
                                                     David J. Goodman

THIRD GENERATION
WIRELESS INFORMATION
NETWORKS

DEVELOPMENTS IN WIRELESS COMMUNICATIONS -- NAVIGATING THE REGULATORY MORASS

By Lawrence J. Movshin
Thelen, Marrin, Johnson & Bridges
805 15th Street, N.W., Suite 900
Washington, D.C. 20005-2207

Abstract

As the United States, and indeed the international communications community, moves toward the development of ubiquitous wireless communications facilities and services, complex and often difficult regulatory issues dealing with the allocation of spectrum and the regulation of the marketplace for these communications networks will be addressed by the Federal Communications Commission, the National Telecommunications and Information Administration, the United States Congress, and even the various and diverse state public utilities commissions. Since the availability of mobile communications will necessarily be spectrum driven, the federal government will need to prioritize the available opportunities in making available new spectrum -- or limiting access to existing spectrum -- for newer technologies. And as wireless communications networks and service providers begin to provide effective competition to the wired telecommunications network, regulators will be faced with difficult choices in assuring universal service at reasonable rates while encouraging advanced communications facilities for their constituents. This paper will explore the state of regulation in the United States in the wireless information arena, looking initially at the activities of the FCC, NTIA, and Congress in allocating spectrum and setting standards for mobile communications networks.

Perhaps no communications-related advances conjure up more vivid images to the typical consumer than the "wireless communicator". Whether they think of the Dick Tracy wristwatch radio of the 1940's, reintroduced to a new generation by Warren Beatty's recent movie, or the ubiquitous flipphone pocket radio into which Captain Kirk utters the famous "Beam me up, Scotty", most Americans have some exotic concepts of how they will participate in the wireless society. Of course, technology advances in the 80's have given most consumers empirical evidence that these comic strip and movie concepts share more truth than fiction -- whereas "radio" communicators were once the province of police departments, truck drivers and taxi cabs, advances in paging and cellular technologies have provided many consumers with a variety of opportunities to use, and realize the substantial benefits of communications mobility.

Added to this mobile communications explosion has been the ubiquitous introduction of information terminals into our every day existence -- from point-of-sale banking to lap top workstations, consumers regularly access user-friendly information gateways which break down many of the historic distinctions between the home, office and even commuter environments. This mobile communications explosion domestically has been further affected by the democratization and industrialization of Eastern Europe and many of the South American countries. These societal changes have spurred increased interest in mobile communications systems, which are seen as providing the greatest potential for introducing advanced communications capabilities to societies which, even today, have only an elementary and antiquated wired telephone network.

In many respects, then, as societies become more mobile, and less oriented to particular fixed locations for particular lifestyle functions, the need to provide a low cost, user friendly, mobile means of communicating -- to cut the umbilical cord of the communications medium -- becomes even more critical to the societal advance. And in that regard, the pressures on technologists to develop such systems, and on the regulatory agencies -- the Federal Communications Commission ("FCC"), the National Telecommunications Industry Association ("NTIA"), and probably even Congress -- to allocate spectrum and develop the regulations necessary to encourage such technology, will be substantial.

It is virtually a given that technologists will be able to improve upon today's wireless capabilities to create better, smaller, more powerful wireless networks and systems which will continue to attract consumer interest. Indeed, demand for more communications mobility assures that technology will achieve such advances. That industry today refers no longer to "wireless" communications, but rather to "tetherless" systems - in order to further establish that no relationship will exist between a "telephone" and a particular base station -- proves how far the concepts have developed even over the last few months. However, because the ability to achieve wireless progress will necessarily depend upon the availability of radio spectrum allocations and/or implementing regulations, I will focus on the impact of those governmental and administrative bodies charged with allocating the use of the radio spectrum -- and the administrative processes which technology pioneers will face in obtaining requisite spectrum allocations and implementing regulations. Before briefly reviewing the various proposals now pending before the FCC and Congress for advancing mobile communications, I will consider the various fora in which decisions will be made, the issues likely to be addressed, the major sectors likely to be involved in the regulatory fray, and likely timetables for decisionmaking, as these agencies work to meet a global society's demands for ubiquitous wireless networks.

Technologists dedicated to improving wireless technology may nevertheless wonder why they require a knowledge of what's going on in Washington, D.C. Simply stated, emerging wireless data communications markets are spectrum driven, and radio spectrum is regulated by the Federal Communications Commission under the oversight of Congress. Will Rogers always advised his friends to "invest in land -- cause they ain't making any more of it." And the same can be said of radio spectrum, the scarcity of which is becoming increasingly evident as a mobile society refuses to be bound by the limits of a telephone wire.

Spectrum scarcity has been evident since the 1970's, when the FCC engaged in its first major reallocation of the UHF Spectrum and allocated major portions of the bands above 800 MHZ to the land mobile community for private trunked and common carrier cellular systems. Today, virtually every new telecommunications advance -- and indeed, even the new forms of mass media distribution like High Definition TV and Direct Broadcast Satellite services -- involves the use of the radio spectrum. Fights for spectrum to satisfy the growing demand for existing services are being fought among the cellular, private radio, mobile satellite and even air-to-ground service providers. There is a snowballing interest in providing ubiquitous Personal Communications services. There is a separate, but certainly related, interest in providing wireless local area networking, both for data and voice, the Wireless PBX. Perhaps no one has better stated the issue than the pundit who predicted that in the next decade the common forms of communications that have previously been provided over the air -- principally radio and television and other forms of mass communication -- will be moved to a wire (principally fiber optics and cable), while the common forms of wired communications, the telecommunications network, will move to the radio spectrum and wireless communications, thereby reversing -- indeed standing on its head -- the communications hierarchy of the last 50 years.

Currently, battle lines for spectrum are being drawn in three distinct fora. First and foremost, the FCC is being challenged to find the spectrum necessary to meet the growing demand for wireless communications. And as the spectrum scarcity starts to impact our nation's leadership in the world communications and information services marketplaces, increased awareness and emphasis is being given to spectrum allocations in the Congress, which is in many respects the final arbiter over the use of the spectrum by the government and private sectors. Ultimately, some of the key battles will be fought on the international front, as various economic sectors seek to achieve spectrum allocations for their own industrialized interests at the 1992 WARC.

One cannot overlook, however, the difficulty of keeping current a discussion of the state of regulatory issues in the incredibly dynamic area of wireless communications. Not a week -- and in some weeks, not a day -- goes by that some new development or some new market entrant, is not announced in the area. The technological solutions and spectrum demands needed to satisfy each market segment may be quite different. Nevertheless, each of these proposals is likely to face a significant and difficult regulatory battle in finding the necessary frequencies. One therefore needs to explore and understand the regulatory process that is likely to drive the government agencies' consideration of new opportunities in the wireless communications marketplace.

One who is not adequately abreast of the regular doings at the FCC and in the Congress will constantly be reacting to, rather than positioned for, the new opportunities in the wireless marketplace. While technologists try to improve spectral efficiency, expand capabilities and decrease the costs of various services, the regulator must continue to balance the resulting, and often competing, demands for spectrum from the various proponents of the services and products which result from the technologists successes; the Hobson's choice which regularly faces those who are charged with assessing competing demands for limited resources. Whether its a city planner who must decide whether to allow expanded development before there are adequate roads and waste treatment facilities, or the environmental planner who must decide between the need for lumber and the need for natural resources to maintain wildlife habitats, the FCC is coninually faced with demands to increase the spectrum available for wireless telecommunications, at the same time that it faces demands for spectrum for new forms and mass media distribution and for spectrum to meet the increased demands of existing service providers for spectrum to meet their own growth projections. The job of the FCC then is to incent both new users and old to improve spectral efficiency through technology, and ultimately to choose those technologies and services by which the maximum number of users and uses can be made of the radio spectrum for purposes that cannot be satisfied through some wired technology.

What then are the processes that the FCC will go through in divining answers to these spectrum allocation questions. Because one party's use of radio spectrum has the potential for interfering with another's, the radio spectrum has long been allocated among potential users, both internationally and domestically, by the governments of the world. As a starting point, spectrum allocation is, quite logically, an international matter, since radiowaves are not sufficiently intelligent to stop at the border of an given nation. Approximately every ten years, a World Administrative Radio Conference ("WARC") is held under the auspices of the International Telecommunications Union ("ITU") at which the nations of the world "divvy up" the use of the spectrum among three different regions -- Europe and Africa (Region 1), North and South America (Region 2) and Asia and the Pacific countries (Region 3). Interim conferences are held to reallocate spectrum among particular services or within particular regions as advances in technology and/or particular unfulfilled needs are identified.

WARCs are extremely significant events -- they effectively design the uses to which radio spectrum can be peacefully put within a region. They also attempt to reconcile the socio-economic needs of various countries within a region with the techno-industrial advances of other more advanced nations. Radio waves cannot easily conform to political boundaries; and yet the intrusion of radio signals can be as threatening to some political regimes as the intrusion of hostile armed forces. The negotiations and decisionmaking at a world or regional Radio Administrative Conference can therefore be as delicate as any diplomatic mission.

The ITU will sponsor a world-wide conference, scheduled for May, 1992, (WARC '92) at which a variety of new spectrum allocations will be considered. It is anticipated that many industrialized nations will propose allocations for new mobile services. The industrialized nations of the world will place great emphasis on the need for adequate mobile spectrum in order to assure the continued development of advanced wireless networks for the movement of information which will be critical to the growth of a globalized economy. The issues being considered at WARC '92 will therefor have substantial ramifications both domestically and internationally for the U.S. information and communications industry.

The FCC has established several informal working committees to study various technical and economic policy issues which are likely to be addressed at the 1992 WARC. Having already requested public comment upon the general areas of concern about which the United States should be most active, the Commission recently issued a further request sharpening the issues as they relate to several possible allocations for Mobile Services.

In the Further Notice, for example, the Commission has tentatively rejected requests that it urge an international exclusive allocation for spectrum in the bands that are likely to be used in Europe for PCNs, refusing to limit its domestic options at this time. The Commission has decided, instead, to urge the maintenance of existing Region 2 co-primary allocations to the Mobile and Fixed Services, leaving to future domestic rulemakings the decision on whether, and in what band, to authorize new or additional spectrum for land mobile services. On the other hand, recognizing the steady growth in the demand for, and proposes uses of, Mobile Satellite Services, the Commission has decided to urge a generic Mobile Satellite allocation, and an allocation of to recognize the needs of emerging Low Earth Orbit Mobile Satellite Systems, while also assuring that other users are not displaced until LEO MSS systems prove out. How well these proposals will be received, and how successful the FCC will be in convincing first the NTIA, and then the rest of the nations at the WARC, to adopt them, will be major policy battles of 1991.

Moreover, since the treaty must be acceptable to all countries, it is more typical than not that if a country has a particular use for a frequency band that is not acceptable world-wide, or even region wide, it will simply reserve its use on a non-interfering or pre-coordinated basis with neighboring countries within the region. The United States, being among the technologically most advanced of nations, has in the past had a large share of the excepted uses. Most of the exceptions could be handled domestically without affecting other countries' use of the spectrum. In those few areas along the Canadian and Mexican border where particular mobile or fixed uses could affect use outside our borders, use was limited to expressly coordinated functions.

However, as communications are globalized in the future, and radio networks are envisioned to be ubiquitous throughout the world, that reservation system may be less effective, and more international compromises may be necessary to allow our domestic telecommunications industry to compete effectively in the provision of both equipment and network services to the global marketplace. In other words, the availability of specific frequency bands to satisfy the United States desire for meeting the requirements of the next generations of wireless communications may be far more dependent on the political vagaries of the WARC negotiations, than on our domestic technological advantages.

Having received the world's intended frequency allocation scheme, the FCC and the NTIA divide up the spectrum between the government and private sectors. Some spectrum is typically allocated exclusively to either side, while a limited amount of spectrum is made available on a shared basis -- primary to one side, but also available on a secondary, non-interfering basis to the other. An Inter-departmental Radio Advisory Committee, representing the NTIA and its constituent agencies and the FCC coordinates such shared uses.

The FCC's decisions on allocating the spectrum reserved for the private sector are governed by the provisions of the Administrative Procedures Act, which require that decisions on spectrum allocation be made only after notice of an intended decision -- and an opportunity to comment -- has been given to the public. In today's highly charged, spectrum scarce environment, it is reasonably certain that any effort to obtain new rules that will create new uses -- or work an effective reallocation -- of any part of the spectrum will invite substantial debate from a wide variety of interest groups intent on protecting what they have, what they intend to use, or what they anticipating needing in the future.

Requests for spectrum come to the Commission in many forms. In some instances, system developers or interest groups will file petitions for rulemak-ing, providing a specific proposal to the Commission for new regulations designed to effect particular uses of the spectrum. In other instances, technology breakthroughs are the result of experimental or developmental licenses issued to spectrum pioneers. When those efforts bear fruit, that fruit is often accompanied by a requirement for a spectrum allocation. And in some cases, counterproposals in one proceeding, which cannot reasonably be satisfied in that case, may lead to the initiation of a distinct proceeding designed to satisfy the needs of the counterproponent.

In some cases, the Commission recognizes the need to satisfy certain societal demands for new services, but the record on what those demands are, and what is the best way to satisfy them is so sparse or so contradictory, that the FCC is not yet sure what to propose in a rulemaking notice. In those cases, it may instead initiate its consideration with a Notice of Inquiry, in which it will pose a series of questions for public comment, often about the state of technology, the potential demand for competing proposals, alternatives available for meeting those demands, and similar general questions designed to elicit broad based responses on a variety of different alternatives. When the issues are truly esoteric, not even adequately definable into questions in a Notice of Inquiry, the FCC will go the advisory committee route, creating a "blue ribbon" panel of industry leaders from the government and private sectors to study the matter and report to the Commission with an adequate record from which it can move the matter forward.

Whether the issues start as a petition for rulemaking, develop from the progress of an experimental license, or are the results of the work of an Advisory Committee or the comments on a Notice of Inquiry, the Commission must issue a Notice of Proposed Rulemaking, in which a specific allocation scheme or technical regulatory restructuring is proposed, before rules are actually adopted. After a record is developed from the public comment on that proposal -- and typically only after months (occasionally years) of further debate and analysis by the Commission and its staff to resolve the often contrary and conflicting views, new regulations and or a new spectrum allocation, will be made. And often those regulations continue under administrative -- and usually judicial -- reconsideration for several years thereafter.

It should be readily apparent that a proposal for a new spectrum allocation -- particularly one intended to satisfy the requirements of a new radio service or new technological breakthroughs -- is not likely to be adopted quickly. Indeed, the Commission's history would suggest that allocation proceedings will run, start to finish, closer to three years than three months.

Obviously, that can be an unacceptably long time for a technological innovator to wait before it even gets the chance to prove out its idea. But the Commission's processes are not stacked entirely against innovation. On the contrary, innovators may apply for an experimental license to build and operate new communications devices and systems in a controlled environment in order to prove out both the technology and the consumer acceptability and usefulness of new products. Cellular technology was developed pursuant to experimental licenses in Chicago and Washington, D.C. in the 1970's and many of the new wireless network technologies are certain to follow a similar route in the 90s.

Experimental licenses can be issued by the FCC's staff; the license will typically grant the licensee the opportunity to develop and even to operate -- on a non-interfering basis -- new technologies and or systems on as large a scale as the licensee can prove is needed to establish the viability of the new equipment or service. The FCC has recently even allowed licensees to market new services or products commercially *in a controlled environment,* in order to establish the demand elasticity for the proposed services. The decade long market test of GTE's Airfone service, and the more recent PCN proposals, all include a "commercial" marketing component. However, as a general rule, the Commission is loathe to allow any wide scale commercial operations which might create substantial consumer expectations even before the new product or service has been permanently authorized spectrum allocation or rules revisions necessary to fulfill its intended scope.

Virtually every Commissioner has at one point in his or her career commented that the most difficult decisions faced were those in which spectrum was being allocated among competing demands.

Commissioners regularly feel like Solomon, being asked to choose between providing adequate spectrum to assure the reasoned and gradual growth of existing services, on the one hand, or encouraging innovation and new services within the available spectrum resources, on the other. It is no wonder that any technology that can promise more uses within less spectrum is so openly welcomed at the Commission, almost without regard to whether such technology has been tried and proven effective.

Moreover, with spectrum allocation proceedings regularly stretching over many years, by the time that decisions on a requested allocation have been made, the original innovator has often long run out of resources necessary to maintain its preeminent technological leadership position in the marketplace. Or larger, well funded organizations have joined the fray, better able to take advantage of the new rules, thus denying to the innovator the bulk of the benefits of his innovation. The FCC has recently recognized the stifling effect that its processes can have on innovation and the ability of smaller companies to raise the capital necessary to create new developments. It has proposed to give a "pioneer's preference" to reward true innovationers with a headstart in the marketplace if an idea is successfully adopted through new rules. But even this proposal has created controversy both as to the definition of what is truly a "pioneering innovation", and as to what the rewards should be to assure that the innovation, in fact, achieves the widest use in the marketplace.

With this introduction to the administrative processes that will be followed in reaching a decision on the allocation of spectrum necessary to cultivate the next generation of wireless networks -- and an understanding of the general pitfalls that are likely to face the proponents of new services designed to take advantage of the various technology advances that will be described throughout this symposium -- a review of several major proceedings that are most likely to result in the availability of new wireless spectrum may provide some insight into the best opportunities for achieving both short term and long term success in bringing the next generation of wireless systems to the marketplace.

With the greater demands in the private sector for spectrum to serve a variety of advanced design, wireless services, and with equally pressing needs to trim the government's own spending for telecommunications research, products and services, the idea of retaining large blocks of spectrum for the government's exclusive use is meeting growing resistance in Congress. The legislature is necessarily concerned not only with cutting government spending but also with finding the resources -- including spectrum -- necessary to allow our domestic telecommunications and informations services industries to thrive in the global market. Probably the best long term solution to the problem of spectrum shortage for private sector initiatives is therefore found in the congressional debate over the Dingell/Markey "Emerging Telecommunications Technologies Act of 1990". The pressures on the FCC to find spectrum for all of the advanced technologies has landed the issue squarely on Congress' lap, resulting in this legislation, passed by the House of Representatives but still stalled in the Senate, which would reallocate up to 200 MHz of radio spectrum from the government use to the private sector, ostensibly to meet growing "high tech" spectrum demands.

The bill would require the Secretary of Commerce within 24 months to identify for reallocation not less than 200 MHz of spectrum below 5 GHz currently assigned to Government use that are not required in the future, and that are most likely to have the greatest potential for commercial uses. The Executive Branch could substitute other frequencies for those proposed in the report, but only to the extent it could justify the change as necessary for the national defense or public safety. One year after the report was approved, the FCC would be required to submit a plan for reallocation of the available spectrum. Not only would the FCC have to designate the intended use of the reallocation, but it would also have to reserve a significant portion of the reallocated spectrum for use not less than *ten years after* the enactment of the Act. In addition, the bill would allow the Executive Branch to reclaim parts of the reallocated spectrum in the future, upon a showing of substantial need. In that case, the government would bear the costs to private users associated with the reclamation of spectrum from the private sector. Finally, the bill provides for the creation of a private sector advisory commission to study, and recommend a reformation of, the processes currently used for allocating the spectrum between the private and government sectors.

Several motives are cited for the push to pass this legislation. On the one hand, the spectrum shortage is sufficiently severe to threaten our domestic industries' ability to maintain a leadership in advanced technologies, with the related concern that other societies will advance further than ours because of the improved communications resources available. It is also recognized by the congressional leadership that the maintenance of a separate, and in some instances, parallel communications system for the exclusive use of the government is a luxury that this nation can no longer afford, except where the nations defense or public health or safety is involved. The privatization of a substantial portion of the government's non-essential telecommunications needs, in order to reduce government investment and spending in this area, has therefore been a strong motivation to finding adequate spectrum for meeting the private sector's needs.

In response to the congressional pressure mounted in favor of the bill, the NTIA has initiated a broad inquiry to reassess how it manages and uses the spectrum allocated to it. At the same time, however, the NTIA and the White House have mounted a strong opposition to the Bill, arguing that the allegedly underutilized spectrum does not exist, and that the Bill's required reallocation will have a devastating effect on the government's -- read military's -- ability to conduct its critical communications functions. Indeed, the White House has threatened a veto of the legislation, and with other communications bills already well ahead of this one, and virtually no time for debate left in this Congressional session, the likelihood of passage this year is slight. On the other hand, the strong Congressional interest in finding relief from the spectrum shortage in the large allocation available for government use has awakened the NTIA to reassess its allocations, and the long-term prospects for finding spectrum for the future generations of wireless communications from this spectrum are good.

It is virtually impossible to get through a week without reading the acronyms CT-2, CT-3, PCN, DECT, GSM, CAI, FDMA, TDMA, CDMA. These are the buzzwords for the concept of a ubiquitous Personal Communications System, in which individuals will be capable of calling, and being called, on a phone identified to them, rather than on a phone identified with a particular station. The term "Personal Communications Services" covers a wide variety of potential offerings. Because cellular service is capable of providing "portable" service, these new ideas have been distinguished from the macro-cell technology of conventional cellular systems by their anticipated use of a "microcellular" design approach.

Unlike conventional cellular and SMR systems, in which large, high powered base stations are designed to cover significant areas, microcellular "personal radio" systems are designed with smaller, lower powered base stations which cover relatively small areas. Whereas cellular systems are designed to provide coverage to fast moving, mobile radios, handing off the caller from cell to cell as he or she moves rapidly through the system, the microcell systems assume largely pedestrian traffic, with the subscriber unit being hand held and typically staying within a cell's range for most of the duration of a call. While there may be some microcell-to-cell hand-off as a pedestrian walks through the system coverage, the number of hand-offs on a particular call should be a fraction of the number in a conventional system. More importantly, the microcells will be closely spaced within a metropolitan area, and less power will be needed by the subscriber unit to reach any given base station. Therefore, the subscriber unit can be comparatively small, using smaller, longer life batteries, and costing much less than the typical cellular portable (which, because it needs greater power capability in order to reach more distant cellular base stations, is typically larger and more expensive than the planned size of the microcellular unit.) This microcellular technology can also be used in an office environment for wireless PBX's and wireless local area network applications. The so-called CT-3 technology is the most recent advance, applying the digital standard DCT-900 to create a wirelss PBX application.

Microcell technologies have been introduced successfully in Great Britain, where they are being used as part of a substantial upgrade to the highly inefficient and antiquated pay telephone network in that country. The most basic microcellular system is the so-called CT-2, advanced cordless telephone network. In a CT-2 system, users will buy or lease a handset that will be utilized for OUTGOING calls only, whenever the subscriber is within a designated distance of a system cell site. These "wireless" payphone base stations will be strategically placed in airports, malls and downtown business centers. When a subscriber desires to make an outbound call, he or she may do so if they are in range of one of these base stations. More advanced system designs may incorporate a paging capability into the handset, so that the subscriber may be paged, and then use the CT-2 network to return the call to the paging party, giving the system the appearance of a two-way network. Because there is no effective payphone network in Great Britain, and the metropolitan area is comparatively small, CT-2 has been popular; whether that popularity will occur in the United States, where there is in fact a very mature and effective public payphone network, and where the number of systems to which a consumer must subscribe, absent nationwide intersystem operatibility, would be enormous, is subject to great debate.

Of greater interest both in Europe and the United States is the use of microcellular technology for full duplex Personal Communications Networks. In such PCN systems, using a concept quite analogous to the cellular system design, the operator will develop a large network of hundreds of microcells within a designated urban or suburban location, each with a limited range of communication. As in the cellular system, the cells will have the ability to hand off callers as they move throughout the system, and/or to maintain balanced capacity within each cell. Consumers would be able to place a call within the coverage range of the network, and also could receive calls wherever they were in the network's coverage. Because the talk-back range to any given cell would be relatively short, a subscriber handset would not need substantial power to communicate within the system -- less than one third of the typical cellular portable unit -- and it could thereby be designed to be much smaller and lighter than the cellular portable, and also much less expensive. And because the cells were closely spaced and with shorter range, frequency reuse and therefor spectrum efficiency would be greater in a micro-cellular PCN, presumably allowing for more subscribers on less spectrum. Ultimately, it is hoped -- if not anticipated -- that individuals would be able to carry their phones, and more importantly, their unique phone numbers, around with them at all times, thereby allowing telecommunications to be based on person to person, rather than station to station, capabilities.

Major policy issues need to be addressed before PCN services can become realities. The Commission is currently considering the broad variety of issues raised by the potential growth of personal communications in a *Notice of Inquiry* in Docket 90-314. Obviously, there is a need to test and prove the effectiveness of the technologies being discussed to digitize the mobile spectrum in a way which would allow microcellular technology to operate effectively in a consumer-oriented telecommunications service. Recently reported studies indicate that up to 40% of the domestic population would purchase and utilize a PCN-type service at various, consumer-oriented pricing levels.

Indeed, the FCC has recently issued a myriad of developmental licenses, with more applications pending, to a variety of applicant proposals to design and test in numerous markets the viability of different types of PCN and CT-2 and CT-3 services and facilities, using various technologies. These developmental proposals will, if effectuated, establish not only whether various technologies can work effectively in urban, suburban and rural environments, but also whether such technologies will allow for the sharing of spectrum with different types of existing users, and also, to a lesser extent, whether there is a substantial consumer interest in particular levels of service. There is a raging debate in the technical community as to whether such ubiquitous services should be accomplished using spread spectrum techniques, which would allow sharing of spectrum with fixed users, or some other multiple access approaches such as Frequency Division or Time Division Multiple Access, which would require a unique allocation of spectrum. These developmental tests may also justify a conclusion to this technological debate.

Even assuming the availability and capability to effectively implement the technologies being considered for microcellular technology, the Commission will have to consider many substantive issues before it can conclude either that a separate frequency allocation is appropriate for PCN and PCN-related services or that some relaxation of technical rules governing various spectrum is necessary to allow PCN services on a shared basis:

- In view of the advanced state of this nation's telecommunications networks, including both the wireline and mobile sectors, is there a need for any or all of the services being considered under the rubric of PCS, or are they instead merely the natural extensions of existing services which will, without further FCC intervention, be developed by existing mobile services providers in the existing mobile spectrum allocation.

* If there is a demonstrated need for any of the PCN services, should the FCC make a specific allocation of primary spectrum for any or all of them, or should such services be accommodated by modifying and liberalizing regulations restricting the uses of and/or technical parameters for existing mobile allocations, as alternative uses of such allocations.

* Should the FCC allocate separate spectrum or make distinct rules for each potential generation of PCN, i.e., CT-2, CT-3, and PCS, or should a single allocation be made for a PCN Service, in which licensees can decide to offer the level of service appropriate to their intended market.

* If an allocation is to be made to PCN services, should it be a primary, co-primary or secondary allocation, and if it is to be primary, what frequencies are available, what will happen to existing users, and how long will any transition off the frequencies by existing users last.

* If an allocation is to be made, should the FCC mandate a particular technology for such services, should specific interface standards be imposed to assure network compatibility, should subscriber equipment protocols be mandated as in cellular, or should each service provider be able to market proprietary customer equipment.

* How should PCN services be regulated, as a common carrier or private radio service; should there be a limit on the number of licensees per market, or should the limit be based on interference criteria; should any segment of the existing mobile industry be prohibited from participating, as local exchange carriers are prohibited from being SMRs, or alternatively, should some PCN or CT-2 licenses be reserved for a particular group of carriers, as one cellular license was reserved for the local exchange carriers.

These issues are all presented for consideration in the *Notice of Inquiry* in Docket 90-314. Initial comments in this proceeding were received in early October, with more than 100 parties filing substantive comments. Not surprisingly, virtually every segment of the communications industry weighed in with an opinion on one or more of these issues. Existing users of the spectrum and existing service providers questioned the wisdom, need and efficiency of a separate allocation of valuable spectrum for a still untried, and questionably valuable service, while the more entrepreneurial groups urged expedited consideration of a separate allocation of anywhere from 140 to 200 MHz for a new PCN radio service, principally in the 1.8-2.1 GHz band. Nor was there unanimity on the technical standards which would govern the use of any allocated spectrum, with proponents of Time Division Multiple Access and Code Division Multiple Access split as to which of these technologies will provide the highest quality, lowest cost, spectrally efficient and greatest capacity for such services.

That there is interest in the development of advanced Personal Communications services and systems cannot be gainsaid, based solely on the size of the response to the *Notice Of Inquiry* in Docket 90-314. On the other hand, the ability to achieve a consensus from such diverse interests, absent a substantial technological breakthrough which would avoid the need for new regulations or allocations, appears achievable only after a long, arduous, and likely rancorous rulemaking proceeding, which is certain to include highly complex social, economic and political issues and require the delicate balancing of significant interests. Such a regulatory structure does not appear likely to provide a short term framework for the commercial development and introduction of new technologies.

Perhaps the best short term solutions exist in two recently completed rulemaking proceedings, FCC Docket 87-389, and FCC Docket 89-354. In the former proceeding, the Commission undertook to rewrite its long-standing Part 15 rules, which govern the use of the radio spectrum on an unlicensed, non-interference basis. Among other improvements in Part 15, the new rules are designed to allow greater flexibility in the use by unlicensed consumer products of certain bands currently allocated for unlicensed Industrial, Medical and Scientific devices. In the latter proceeding, the Commission substantially expanded the technical flexibility allowed in the use of spread spectrum technologies for unlicensed communications systems. In each instance, after substantial comment and review, the Commission chose to liberalize technical restrictions on the use of bands allocated to other, primary licensed users, in order to allow greater unlicensed use, *but only so long as such use does not interfere with the primary licensees.*

Using the technical flexibility provided by the newly adopted rules, a number of companies have designed products which can work effectively in a low power, microcellular environment, not only without creating interference, but also without being susceptible to the relatively stronger signals of the primary licensed services. Recently announced CT-2 and Wireless LAN products have been designed to utilize the benefits of Part 15.

By avoiding the need for a spectrum allocation, many of the issues involved in the prioritization of demand have been avoided. While there remains the necessary demonstration that the shared use of given spectrum can, indeed, occur without creating interference to existing and future users, the evidentiary showing in such case remains technical, and less economic or political. Of course, Part 15 is not without its limitations, principal of which is the lack of any protection from interference from other devices. But this may be an easier technical hurdle to overcome than the regulatory hurdles necessary to achieve an allocation in which systems are protected.

The Commission necessarily welcomes any effort to achieve greater user density throughout the available and usable spectrum. Proposals designed to take advantage of technologies which have the potential for efficient spectrum sharing without the need for a new spectrum allocation necessarily provide the greatest short term potential for achieving commercial implementation of a new generation of wireless networks.

Finally, it is worth reviewing one other regulatory alternative for meeting spectrum demands -- moving to liberalize the regulations governing the use of an existing spectrum allocation so that it can be used in a different, unanticipated, but nevertheless, forward looking fashion. Two examples of this approach are exemplary. Recently, for example, the FCC amended a number of its rules governing the use of the 18 GHz band which is allocated for point-to-multipoint Digital Termination microwave licensees. The new rules remove requirements that these channels be used only for high powered systems, and instead allow their use in low powered, multiple transmitter systems, with much shorter co-channel interference protection radii. At the same time, the Commission eased the prior coordination and application requirements for these services. Taking advantage of these more liberal rules, manufacturers have targeted the 18 GHz band for the development of wireless LANs and PBXs, using the line of sight nature of the frequency band, the lower power now allowed, and the potential in-building and building to building reuse of the substantial number of channels likely to be available in this band to create an entirely new use of otherwise under-utilized channels.

Using a similar tactic, Fleetcall, Inc. recently petitioned the Commission for rules waivers which would allow it to use its Specialized Mobile Radio licenses in the 800 and 900 MHz band, currently limited to higher power, single base station service, in a lower power, multi base station cellular-type system. Fleetcall's major argument in favor of such relief is that the Commission will be able to expand the capacity and use of the channels allocated to SMRs by 3-6 fold over the maximum effectively available under existing restrictions, *without* allocating any new frequencies to the SMR service.

The appeal of providing capacity for the future growth of a vital and dynamic radio service *without* any new spectrum allocation is a strong one. It will, of course, have to be balanced against the argument that the service to be provided by the proposed Fleetcall system no longer meets the needs characterized by the Commission in creating the SMR service, and indeed is far closer to the services being offered by the cellular licensees, only not as efficiently.

In both of these cases, proponents of change looked to re-mold the technical characteristics of an existing service to achieve the capabilities of new technologies instead of looking for new spectrum for an advanced service -- which is likely to require the displacement or inconvenience of existing users. Whether it is the redefinition of the point-to-point microwave service to include low-power, in-building type wireless Lan and PBX applications, or the removal of restrictions on the Private Land Mobile SMR service to allow it to break out of its limited capacity restraints to become more spectrally efficient, each of these ideas allows for the use of existing regulatory and spectrum frameworks to develop new regulatory solutions, rather than looking to move into new, uncharted spectrum turf where the number of potentially affected claimants is necessarily greater. While necessarily time consuming, this approach may be a more efficient short term solution than the major spectrum battle which almost certainly looms for the development of a separate PCS radio service.

In conclusion, it cannot be denied that technology is capable of driving the movement to a wireless society, allowing the consumer to access the underlying telecommunications network on a personalized basis without the limitation of a wire or cable. As the global economy demands the internationalization of information services, and access thereto, the international standardization of wireless technologies and services will become more important to the global economy, and the United States' role in it. Therefore, if the limits of technological innovation are to be achieved in this country, the federal regulatory regime -- from Congress through the executive branch and into the Federal Communications Commission -- will play a key role in determining not only what technologies are favored, but also what markets are served, and how.

Technologies that can share spectrum with existing users or substantially increase the capacity of existing services will certainly be favored over less efficient proposals. The FCC will likely be forced to deny spectrum to certain uses in order to provide the most economically and politically favored users adequate spectrum for meeting anticipated demands. While European models are likely to hold some influence over the decisionmaking -- and domestic manufacturers and service providers will be favored where allocations can make them more effective competitors on a world-wide basis -- the domestic needs of consumers and industry will ultimately be the model used to determine allocations.

The technologists role will be to continue developing greater spectral efficiencies into devices at lower costs, so that manufacturers and users are incented to utilize such devices and systems. The regulators role will be to accommodate those systems and services that achieve the greatest spectral densities for the most valued uses. The likely confluence of private and common carriage, of personal and business communications, and of wired and wireless networks into a seamless ubiquitous system, will challenge the limits of both the technologists and regulators creativity in satisfying the public's demand for advanced communications capabilities.

Why do we need standards for mobile/wireless communications?

Dr. Shila Heeralall
Bell Cellular Inc., 20 Carlson Court,
Etobicoke, Ontario, CANADA M9W 6V4

1. Introduction

This paper attempts to put in a single document a comprehensive discussion of the subject of why we need to have standards for wireless communications (mobile and wireless[1] are used interchangeably in this paper). The paper also looks at the effectiveness of our present arrangements to protect patents and foster innovation. Much of what is said is true for any industry. But telecommunications is not quite like all other industries. It deserves special consideration because of the fact that its networks must interconnect to provide basic service. Furthermore, in mobile communications users should be able to roam as widely as possible.

The subject of this paper is a multidisciplinary one (law, sociology, economics, management, national psychology and corporate psychology). It can be a very interesting research topic. I hope I can at least stimulate some discussion by presenting my own views here.

2. Why do we have regional standards for mobile communications?

Regional community of interest is already a recognised fact. That means customers expect regional roaming to be possible. Indeed the value of a mobile phone to its user increases significantly when it can be used in more places. Manufacturers for

---

[1] mobile communication: radio communication with continuous mobility through many cells during the call, i.e. with handover from cell to cell.
wireless communication: radio communication without handover from cell to cell during the call.

their part can achieve lower production costs if there is one standard for several markets than if each market has a different standard. Regional free trade blocks such as Europe and North America are adding to the spirit of 'effective competition through harmonised standards'. Customers within the region benefit from: the roaming convenience, the competitive forces as well as the economies of scale.

Unfortunately, sometimes a regional standard is regarded as a non-tariff barrier against foreign competition or as an assertion of the region's identity. Sometimes it is argued that the regional standard is designed to meet specific regional needs. Such reasons will no longer be relevant in the future due to increasing globalisation.

3. Intellectual property rights today

New ideas and innovations are protected by patents and only the patentees have the rights to use them and/or licence them. The patents are mostly owned by companies, although individuals can also be patentees. Some patents are filed in several countries for international protection. The reward to a patentee is either royalties from those that use the patent or competitive advantage through exclusive use of the patent. A patentee may decide to make certain patents available royalty-free in return for some indirect commercial advantage. Unauthorised use of a patent can be costly to those infringing the law.

When a patented idea can only be used as part of an industry standard a dilemma arises. How can the developers, usually manufacturers, recover their development costs without claiming royalties or how can they be exclusive while offering a proposal for a standard? What is the reward for being innovative? Consumers would like to benefit from all innovations irrespective of their sources. Presently this is not always possible. Standards bodies everywhere have not found the ideal formula yet. Meanwhile it is still difficult to make patented ideas become standards.

4. Motivations to innovate today

Companies have a strong motivation to innovate in areas where they have the prospect of exclusive rewards through patents. Employees get various incentive packages such as bonus and/or profit sharing. A private innovator can cash all the reward alone, but is deprived of team enrichment of his ideas. Even within companies it is a challenge to optimise the individual and team contributions. Within a standards body, the companies are like the individuals who together constitute a team to serve public interest. Yet we have no mechanism to always bring out the best proposals for a standard. Commercial insecurities can create a stalemate or lead to reluctance to adopt standards. What is missing perhaps is the reward to individual companies. The victims of lack of concensus on standards are consumers. It is popular for businesses to claim to be customer oriented. Yet, how much do consumers get a chance to express their wishes? They can choose from the options given to them, but they cannot tell what options they really want. Innovations do not always match what consumers want.

5. The telecommunications industry, a special case

The discussion so far could apply to many industries. But the telecommunications industry is a special case. It is one in which it is particularly difficult to bypass standards. Both compatibility and co-existence standards are needed. This is because users must be able to call each other across the world and across competing networks. So all networks (regional or competing) must interconnect at some point to form a super network - the global PSTN. Furthermore, it is necessary to have a high degree of organisation and concensus among operators of the global network. Signalling translations and other conversions take place at gate exchanges. End-to-end quality is controlled by determining what maximum amount of degradation should be tolerated in each part of the network. The ITU plays a crucial role in making the global network possible.

By contrast, in the car industry, standards exist to ensure that public interests are above commercial interests e.g. gas exhausts, passenger safety features. But different cars do not need to interact, they only need to run on the same roads.

6. Furthermore, the special case of mobile communications

In section 2, I explained some of the reasons why there are regional standards for mobile communications. As the global network evolves to include mobility and personal communications, there is a growing demand for international roaming. People expect to be able to use the same terminal wherever they go. This means standards for mobile communications will eventually have to be harmonised across all regions of the world. A global mobile network requires even more international standards than the global PSTN, thus making mobile communications a further special case within the telecommunications industry. For example, universal access can only be achieved by agreeing on certain standards. Such standards include signalling protocols on the radio interfaces, and technical specifications for customer equipment and base stations. International call delivery requires mobility management on a global scale.

7. Mobile communications in the 80's and 1990

In telecommunications, the 80's has seen the start of mobile communications for the masses. Although cellular is the dominant one, there has been a trend towards several mobile communications services: analogue and digital cellular, digital cordless telephone, PCN's[2], mobile data, paging, skyphone, dispatch, mobile satellite. There is confusion because of the rapid evolution and introduction of new services together with increasing deregulation of service provision. Services cannot yet be integrated technically or commercially and there are no international standards. Furthermore, the ultimate role of the PSTN is not clear.

---

[2] PCN's Personal Communication Networks

Regarding standards, the number wortldwide for cellular has decreased from several for analogue to three for digital. That decrease is mainly due to the harmonisation efforts in Europe to achieve pan-European standards for digital cellular. Standards for digital cordless telephony have started appearing in the late 80's and the situation is still evolving. Standards must increasingly facilitate competing networks to co-exist and interwork. Since 1985 CCIR IWP 8/13[3] is working towards a single international standard for FPLMTS[4]. That will also provide for integration of services. Meanwhile an increasing number of international consortia are helping to blend regional business interests and the idea of integration is everywhere. More recently, PSTN's are also showing keen interest in personal communications. So CCITT and CCIR are trying to harmonise their efforts[5].

8. The 90's and 21st century

The 90's and early 21st century are being hailed as the age of the global village, environmental concerns and personal communications among other things. Futurists[6] are talking about free trade among all nations, global economy and one marketplace. They say telecommunications made the global economy possible in the first place. My view is that in the 21st century international mobile communications will promote global harmony and global free trade. Two billion passengers will travel the world's airways in year 2000. We will disappoint them if we do not provide them with the freedom to keep in touch as they go about. In the developed countries people are returning to rural communities but these will have urban communication facilities (smart houses, home offices). In developing countries people are still converging to the cities and are still in need of basic communication facilities to develop their economies. It is said that global warming will create new migration patterns. Global telecommunication standards of the future will have to meet the needs of not only a highly mobile society but also urban and rural communities and migrating communities.

9. 21st century values

Globalisation by itself introduces an international spirit. In addition to that, new problems such as global warming and air quality are forcing nations to collaborate in more socialist ways for their common good. Such collaborative spirit together with the continuing wave of privatization changes the meaning of economic interest. That may lead to a more abstract meaning of competition and how it should be made constructive to serve consumer interests. Perhaps competitive advantage will come through genuine concern for the impact on the community of one's products and

---

[3] CCIR IWP 8/13 Interim Working Party 13 of Study Group 8 of the International Consultative Committee for Radiocommunications. Its mandate is to study FPLMTS

[4] FPLMTS Future Public Land Mobile Telecommunication Systems

[5] A Joint Experts Working Group meeting of CCITT and CCIR took place in Vancouver, Canada in May 1990.

[6] "Megatrends 2000 - Ten New Directions for the 1990's" by John Naisbitt & Patricia Aburdene, William Morrow and Company Inc. New York, 1990, Chapter 1.

services. At present, competition can sometimes be against consumer interests, e.g. when there is reluctance to agree on standards.

In the future, individuals will have more opportunities to make a difference[7]. So, in businesses customers will probably participate more in company decisions and industry standards. With an increasingly enlightened public this is quite possible.

10. Why do we need standards for mobile communications in the 90's and 21st century?

Amidst all the new trends, technology keeps advancing at an increasing pace while attitudes and values are changing. At this point in time many of us are reassessing the benefits of having standards for mobile communications. The dilemma is between meeting the needs of the mobile communications industry and allowing innovation to happen without hindrance.

Interests of various categories of people

The subject of standards is of interest to all of us whether we are in the category of service providers, manufacturers, researchers, policy makers, regulators or merely consumers. Mixed views as to the benefits of having standards is the result of (seemingly?) conflicting interests from different categories of people. Consumers, policy makers and regulators want high quality products and services at competitive prices and available to most/all people. They usually welcome standards. Researchers and manufacturers may sometimes consider standards as a hindrance to 'any time' innovation. But in practice service providers and customers do not welcome chaotic evolution. Business plans are upset and customers pay the price. Since all research must be financed somehow, it is better to have orderly innovation through standards. The standards bodies for their part must be responsive in bringing the benefits of innovation to consumers as soon a spossible. Ultimately our common interest is to serve consumers well since without them none of us would exist.

The reasons why standards are indeed needed

Some reasons have already been given in previous sections: global interconnection and roaming, end-to-end quality control, economies of scale, effective competition in free trade blocks. These reasons will still be valid in the future. Some additional reasons are now given in terms of futuristic thinking:

1. No one has excellence in everything. Standards allow the best from each contributor to be put together to achieve higher performances for the benefit of consumers. This is clearly constructive competition at the level of setting standards. Such a framework also allows small companies or even individuals to make meaningful contributions.

---

[7] ditto, Chapter 10.

2.   Customers want effective competition i.e. highest reliability and lowest cost for basic products and services. Likewise service providers want the most competitive products from manufacturers.

3.   Developing countries and smaller markets who typically do not have domestic manufacturers have even more need for an open choice of suppliers.

4.   Effective competition means open choice of suppliers of equipment and open choice of service providers. Open choice means complying to same standards. Excessive proprietary specifications create captive markets primarily for the benefit of manufacturers.

5.   The need to synchronise evolution and introduction of innovations to avoid inconvenience to the customer base.

11. What constitutes standards and who should develop them?

Some of the things that need to be standardised in order to meet the objectives of having standards include:

1.   The radio interfaces - speech coding, modulation, access scheme, signalling protocols and frequency allocations. Regarding frequency allocations, the outcome of WARC'92[8] is particularly important.
2.   Interfaces between major network components such as cell site-to-switch, switch-to-switch and links to data bases.
3.   Interconnection of networks.
4.   Integration of services, this is an important consideration to achieve personal communications.
5.   The human to terminal interface, this is one which deserves more attention for the benefit of users.
6.   Operational aspects such as billing.

Standards should be developed by all those that will use them. At least all those concerned should have the opportunity to participate. This is because all companies/countries need a sense of fulfilment, just like individuals do. International standards must benefit from the participation of all countries/regions in whatever way that it is convenient for them to bring in their contributions. Selection of a standard from multiple proposals should be made by unbiassed scientific evaluations, to the extent possible. Both technical and non-technical evaluations are needed. An example of the latter is marketing. Selections are already being made at regional levels, this is constructive competition.

---

8 WARC'92 The World Administrative Radio Conference will meet in 1992 to look at spectrum between 0.5 and 3.0 GHz.

12. How to stimulate innovation in the 21st century?

As mentioned already, countries and companies are like individuals, they all need a sense of achievement, something to be proud of, and opportunities for fulfilment. They also need to be complimented and rewarded for good work. A merit system is needed to promote quality, efficiency and creativity. Competition should serve both corporate self-interest and public interest. Every time standards are set or evolved, participants should be eager to submit their best proposals and have them adopted, if chosen. Individual patent royalties may not be the best reward for winning proposals. Perhaps a system of awards together with a common pool of royalties shared among the winners would be easier to manage and yet be motivating. There would thus be much opportunity for self-advertising and the associated commercial advantage. Obviously the details of any new arrangement would have to be worked out carefully. Intra-corporate cultures and inter-corporate relationships will surely have to adjust to the new values.

13. Intellectual Property Rights in the 21st century?

The case of Intellectual Properties (patents and copyrights) that are proposed for standards will have to be developed as a special category . This will greatly facilitate and speed up the implementation of standards without sacrificing its quality. Other Intellectual Properties can still be protected according to established laws.

14. Rigid vs. flexible standards

The question of how far standards should be rigid or flexible affects opportunities for innovation and competitive advantage as well as consumer interests.

To the extent that mobile communications is a special industry a significant degree of rigidity is needed to permit interworking and roaming. But rigidity also forces competition on the basis of cost, reliability, system engineering and such things, within a free trade/open competition framework - i.e. no captive markets. Having an open choice of supplier is key to effective competition, it means that trying an alternative supplier does not require a complete system replacement. Since the potential market with rigid standards is very large, consumers have the additional benefit of low cost through mass production - i.e. economies of scale. And competitive manufacturers make maximum sales.

The negative side of rigidity however is that it limits the scope for innovation before the next review of standards. However a well thought out standard can have built into it as much flexibility as desired to allow for the excitement of being creative and exclusive. Anything that does not encroach on transparent interworking or roaming can be left unspecified. Flexibility very likely means an expanded set of signalling protocols and therefore extra software and memory. Depending what options these protocols allow for, there may be other extras. In order to be acceptable to consumers these extras should bring commensurate benefits beyond the basic services. Excessive flexibility would amount to no standards. But certain forms of flexibility may simply be needed by the system, e.g. radio intefaces.I believe that having modular signalling and hardware is a very smart way to be flexible. Modularity also facilitates and speeds up maintenance and allows a variety of market needs to be met cost-effectively. Hardware duplication due to flexibility of standards is perhaps the least acceptable of all extras, e.g. flexible speech coding techniques is too radical but variable bit rate speech coding using the same technique can be very attractive since consumers may have different willingness to pay for quality. Two other significant advantages of flexibility/modularity are: easy customisation to market segments/niches and future proofing of standards.

It is important to strike a balance between rigidity and flexibility for the benefit of everyone.

15. Conclusions

Standards for mobile communications are needed because of the special nature of that industry: its needs for global interconnection, integration of services, roaming across regions and even across the world. But in addition standards lead to effective competition, economies of scale, higher performances and synchronisation of evolution across the industry. Standards mean better satisfied customers and therefore higher penetrations of service. In the future, globalisation of the economy and increased global mobility will require international standards. Free trade is a trend against local standards.

International standards must address: the radio interfaces including frequency allocations, interfaces between major network components, the human-terminal interface, network-to-network interface and operational aspects.

In order to develop high quality standards some new ways of rewarding the best proposals are needed. This is to promote competition at the level of standards and get the best outcome. The system of reward must be manageable to facilitate the implementation of standards. Timely reviews of standards is the way to bring innovations to customers quickly.

A reasonable degree of flexibility and modularity provide opportunities for competitive advantage as well as for customisation to individual markets or market segments. The inherent futureproofing is also appealing.

STANDARDS FOR GLOBAL PERSONAL COMMUNICATIONS SERVICES

Mike Callendar
(Chairman CCIR Task Group 8/1, formerly IWP8/13)
MPR Teltech Limited,
8999 Nelson Way, Burnaby,
British Columbia, V5A 4B5, Canada

Abstract

*Global Personal Telecommunications represents a logical evolution
from many of today's radio-based regional mobile services such as
cellular, cordless phones, and paging.*

*Extension of both voice and data telecommunications to a person rather
than a place will have a significant effect on the way we live, and on the
organizations which currently provide these services.*

*The paper outlines some of the international standards work in this area
currently being carried out by CCIR/CCITT, in particular that of CCIR
Task Group 8/1 (formerly IWP8/13) on Future Public Land Mobile
Telecommunication Systems (FPLMTS).*

*The decision to hold a World Administrative Radio Conference (WARC)
in 1992, to consider allocations in the 0.5 - 3 GHz band for the land
mobile and mobile satellite services, will provide a global focus for
world-wide personal telecommunication services.*

Introduction

Many incompatible analogue cellular mobile systems operate in different regions throughout the world but none presently come close to providing truly personal service.

The Telecommunications Industry Association (TIA) has specified a new digital cellular standard to replace the AMPS analogue one used throughout North America today, new standards have also been specified for a Pan-European digital cellular system [1] and a Japanese digital cellular system all to begin operation in the early 1990's. These incompatible new regional digital TDMA cellular systems, while being a considerable step forward will still have to serve both vehicular and hand-held units, and so will likely only go part of the way towards meeting the future needs of Personal Telecommunications (PT).

Trials of several Spread Spectrum (CDMA) wireless access systems are currently planned in North America. The results of these tests will provide valuable input to the standardization process for wireless PT.

A number of new improved digital **cordless phone** designs have been proposed for business and residential use which are also potentially able to originate calls through "**Telepoints**" in many public areas as well. Telepoints will require billing procedures and some level of radio air-interface standardization.

Calls to cordless phones as people roam between home, office and public areas could require an infrastructure comprising location data bases etc., much as in cellular systems. An alternative approach would be to integrate this service with wide-area paging. [2][3][4][5]

Bellcore have proposed a wireless local loop concept which provides Universal Digital Portable Communications by using the ever increasing distributed intelligence of the fixed network to control routing of calls to and from the various low power radio ports.[6]

A consortium of over 20 companies and research organizations in Europe is working on the requirements to provide Universal Mobile Telecommunication Service (UMTS) as part of the Research into Advanced Communications in Europe (RACE) program. The system would consist of a federation of individual mobile and cordless services all working to a common standard. The key elements are standardization of a flexible high-bit-rate radio interface and the provision of supporting functions within the public fixed network. [7]

The increasing technical sophistication of newer digital cordless products, and improvements in the personalization of cellular service, will progressively reduce the distinction between these two markets and could with the necessary international cooperation result in evolution towards a common world-wide standard for wireless PT.

Mobile satellite systems are also being planned to provide coverage in the many areas where terrestrial coverage is not economically feasible.

The **1992 WARC**, which will consider spectrum allocations in the 0.5 - 3 GHz band for the land mobile (**FPLMTS**) and mobile satellite services, provides a unique opportunity to develop a truely ubiquitous PT service.

It is important that an integrated approach is taken to all mobile services which could potentially provide communications with people on the move. This includes terrestrial and satellite based systems serving land sea and air transportation. [8][9]

Personal Telecommunications (PT)

There have been many different visions of future personal and mobile telecommunications [2][3][4][5][6][7][8][9][10][11][12] [13]14][15][16].

In spite of the many variations some common factors emerge:

● PT is a service concept, which will evolve mainly from existing radio-based services such as cellular, cordless, paging etc., and will use existing and planned fixed and mobile infrastructures to the greatest extent practicable.

● The majority of PT will eventually have a wireless rather than a wired link to the person.

● These wireless systems will all involve large numbers of small cells serving economical and convenient low power hand portable units.

● The service quality should become, wherever possible, comparable to that of existing wired services, e.g. for wireless this implies better than 99% coverage with less than 1% blocking.

● Radio system architectures significantly different to today's cellular systems will be required, with an increasing trend towards adaptive control techniques and distributed intelligence.

A PT service requires unique user identification numbers, and associated data bases containing the current routing information, which would replace the several different telephone numbers that each of us uses as we move about today.

The concept of a service conferring mobility is applicable whether a user is served by a mobile or fixed network. The terms **continuous** and **discrete** mobility have been defined to differentiate between what could be achieved by wireless and wired networks. **Continuous mobility** will in general be constrained depending on the type of wireless terminal used to access the service.

A number of potentially different public mobile radio interfaces, operating in various frequency bands, have been identified by **Task Group 8/1,** e.g. indoor and outdoor microcell, macrocell, very large cell, and satellite. "Macrocell" for example corresponds with today's cellular systems, and "very large cell" with terrestrial-based air-ground systems.

While many of these radio interfaces are likely to remain standardized on only a national or regional basis, the final low power radio link to the person could potentially be the same on a global basis.

A Personal Telecommunications Number (PTN) service, which involves personalizing a wired (or even wireless) phone in each location as one moves around, provides flexibility of access or discrete mobility and represents the wired network's attempt to provide personal telecommunications service. This approach would, of course, require data bases containing current location and routing information.

FPLMTS, and other mobile radio systems, can also benefit from this flexibility of access concept. These "higher layers" of service (mobility, flexibility of access) are offered by FPLMTS in addition to the conventional telecommunication services.

The differences from a user perspective between the services provided by the mobile and fixed networks will become less into the future. It will become increasingly difficult for users to differentiate between these two networks.

PT involves providing a largely transparent network linking the various possible wired or wireless bearers so that a practical range of telecommunications services can be provided to people on the move, wherever they may be. **In order to make this potentially intrusive service acceptable to customers it is desirable that control of communications be given to the recipient rather than the originator of the call.**

In current public telecommunication systems, the alerting function always resides in the same terminal used to answer the call. However, in the FPLMTS, it is envisaged that the device on which the alert is received, e.g. pager, personal station (PS). etc., is not necessarily the one used to answer the call. The called party will be able to use any terminal of his choice (e.g. telephone or mobile station (MS/PS) to answer the incoming call. This implies that delivery of a signal to an alerting device is not a complete activity, rather, just a part of the total activity associated with establishing a call.

Separating alerting from answering and future call screening techniques could have implications on the sequence of call-establishment signals, call completion time delays and charging requirements, and need further study by a wide range of experts.

It is likely that, because of limitations inherent in both wired and wireless personal service, customers will need to move between these systems. **There is therefore a requirement for an integrated design approach to all aspects, e.g. wired/wireless, of PT service development.**

Fixed Applications of Wireless Access

Radio-based wireless access to the fixed telecommunications networks is predicted to become progressively more cost competitive with the various wired alternatives. It is therefore likely that an increasing percentage of fixed traffic will originate and/or terminate in some form of wireless device.

Since phone service penetration in developed countries is close to 100% it is easy to forget that **two thirds of the world's population does not even have access to a telephone.**

Telecommunications play an important role in economic development and investments in wireless telecommunications infrastructures throughout the world can be expected to grow dramatically as this technology provides an increasingly cost effective solution to this important global requirement. [17][18][19]

Personal Telecommunications Service Requirements

Extensive cellular radio networks are in operation throughout the world serving many millions of users. The predominant service is voice but data related services are growing in popularity. The planned reorganization of many of the fixed telecommunication networks throughout the world into Integrated Services Digital Networks (ISDN) has focused attention on how best to extend these types of services to mobile users. [20]

Future mobile systems will be expected to provide digital services covering a wide range of bit rates and error ratios. The maximum usable gross channel bit rate is usually limited by multipath radio transmission characteristics which vary by orders of magnitude between small indoor microcells and large outdoor macrocells.

It would be desirable to be able to offer higher bit rate services in areas where radio propagation conditions are suitable rather than be limited to bit rates associated with worst case conditions.

This suggests that the gross channel bit rate, and hence transmitted bandwidth, should be adapted in relation to the size and propagation characteristics of the cell, much as power levels are adapted in today's cellular systems. [21] Baseband digital signal processor (DSP) chips could readily make these adjustments, essentially equivalent to a change of clock rate, as required under system control.

The future range and combination of both packet and circuit switched services required at any instant in a given cell is expected to vary considerably, and hence an efficient adaptable access and control scheme is needed to optimize the available transmission resources. [22]

Wireless extensions of a flexible primary rate ISDN interface could be a valuable future system feature for microcellular radio environments.

Suitable control features should be built into future mobile ISDN's so that the various integrated services can be efficiently queued under busy conditions to optimize the grade of service to all users, under varying combinations of service requirements. This should include the ability to modify the services offered under peak loading conditions.

Acceptable response times for various voice and data services will vary from less than 1 second in the case of interactive data communications to many minutes for some status reporting traffic.

Three broad categories of traffic can thus be identified: voice, time-sensitive data and non time-sensitive data. This suggests that for optimum smoothing of the load within an integrated voice and data network three separate queuing and prioritizing strategies should be considered.

Flexible Standards
The challenge in any standards work is to define things completely enough for reliable operation without restricting future enhancements that can not be specified at the time.

The key to ensuring future enhancements is to consider that some parameters can be variables under system control rather than fixed values defined in the specification. All that needs to be defined in the specification is the control mechanism and the range of variation available.

Although it may not be possible, at present, to build equipment which can readily adapt to operate over the full range of the various parameter variations allowed in the specification, it is still valuable to have this flexibility available to allow manufacturers and service providers future scope for innovative ideas which can not perhaps be implemented now.

It may not be economically practical to use the traditional approach to cellular engineering for future microcellular personal communications systems. [23] More intelligence and autonomy will clearly be needed at base and mobile stations.

Mobile radio systems today tend to be designed with fixed parameters which are usually determined by worst case conditions, this results in reduced peak capacity and better than required performance for most of the users most of the time. Significant overall system improvements can be obtained, for example, if every user is given just the quality required and no more. [24][25]

Adaptive schemes have the potential to significantly increase the capacity and/or perform-ance of future wireless access systems at little increase in cost or complexity.

For example duplex telephone circuits have a voice activity level of around 40%, since both parties do not usually talk simultaneously. If this property could be effectively used, perhaps only under "busy-hour" conditions, it would be equivalent to more than doubling system peak capacity.

Digital Speech Interpolation techniques work well in large trunk groups, as in the fixed telecommunications networks, but it is necessary to add an adaptive voice coding rate capability to deal effectively with the more significant peak activity periods in the small trunk groups common in radio systems. [26]

Hand-held portable units will become a major factor in future mobile systems (possibly 50% of all mobiles by 1995) [23] and so allowance for the significantly different requirements of these units, compared to vehicular mobiles, must be taken into account. Of particular concern are the reduced transmitted power available, the need to conserve standby power consumption and the potential for long fades due to the quasi-stationary use of these units.

Long fades, which could cause a number of successive speech frames to be corrupted, can not be easily dealt with by forward error correction or frame substitution techniques. Slow frequency hopping, diversity reception and even just switching a single receiver between two suitably spaced antennas have been proposed. [27]

Two major digital wireless access technologies, cellular and cordless, will be extensively deployed in the 1990's and will to some extent compete for the same customers. A specification which allows adaptation to cover both requirements would be a very attractive proposition. The increasing availability of software controlled signal processors means that the incremental cost of this increase in flexibility can be very small, while the benefits in improved system efficiency and reduced service costs can be very large.

International Standards Activities

The International Telecommunication Union (ITU), which is responsible for telecommuni-cations standards and spectrum management world-wide, has both regulatory and technical organizations. In the regulatory area the International Frequency Registration Board (IFRB) keeps track of frequency assignments that have international significance, and World Ad-ministrative Radio Conferences (WARC's) are organized to update the Radio Regulations, thus sharing the radio spectrum between the many competing services and administrations. The International Telegraph and Telephone Consultative Committee (CCITT) advises the ITU on "wired" telecommunications requirements, and the International Radio Consultative Committee (CCIR) advises on radio related, i.e. "wireless", requirements.

Many Administrations and Regional organizations are currently studying the requirements for telecommunications with people on the move, including a growing focus on PT. In order to avoid solutions that are limited in scope to regional areas and conditions it is important that studies on an international scale be rapidly carried out in a single forum. CCIR and CCITT represent this forum in the framework of the ITU.

In late 1985 **CCIR Study Group 8**, which is responsible for all mobile services, formed a special international group to identify the requirements for Future Public Land Mobile Telecommunication Systems (**FPLMTS**). 29 Administrations and 10 International Organizations/RPOA/SIO's presently participate in the work of this group, which was called Interim Working Party 8/13 and is now known as **Task Group 8/1.**

In April 1988 a new Interim Working Party (**IWP8/14**) was formed to deal with Mobile Satellite requirements for land sea and air services.

The CCIR Plenary in May 1990 made a number of changes in its study group structure and working methods. Study Group 8 will now have four permanent Working Parties 8A, 8B, 8C and 8D and IWP8/13 becomes Task Group 8/1 reporting directly to Study Group 8.

CCITT, through its various study groups, has developed many mobile-related recommendations in such areas as numbering plans, location registration procedures and signalling protocols.

It is likely that the majority of PT will eventually originate and/or terminate in some form of wireless device. It is therefore logical for the fixed network to provide many of the supporting functions for FPLMTS and for fixed network standards to be adapted to the needs of emerging wireless access systems.

A Joint CCIR/CCITT Experts Working Meeting on PT was held in Vancouver B.C. Canada June 4-8 1990, bringing together experts from Task Group 8/1 and most affected CCITT Study Groups. This unique meeting was held to help coordinate the various wired and wireless standards activities needed to provide seamless PT.

Since most of the future standards work related to the fixed network will be done by the CCITT, and a number of CCITT Study Groups are already investigating PT requirements, **it is very important that close collaboration is maintained between the various CCIR and CCITT study groups.**

The application of mobile radio techniques to the provision of service to fixed locations, usually where there is no existing wired infrastructure, is within the scope of Task Group 81.

This is the result of requests to CCIR from developing countries for study of how these mobile systems may be modified to meet their needs.

Task Group 8/1 Progress to Date

Six meetings have been held so far and have produced a Recommendation and Report on FPLMTS, plus a separate Report on adaptation of mobile technologies to the needs of developing countries. These documents were accepted by CCIR Study Group 8 at its meeting in October 1989 and revised terms of reference for the future work of Task Group 8/1 were also defined.

The sixth meeting, in the UK July 3-12 1990, prepared two additional reports on the spectrum needs of FPLMTS and on sharing considerations between FPLMTS and other services. These will be submitted to IWP8/15, to coordinate with other Study Group 8 material, and to JIWP-WARC-92 which is preparing the technical basis for WARC-92.

FPLMTS involves more than just personal telecommunications, since a flexible digital radio link can obviously be used as a wireless local or wide area network for communications between machines, e.g. for vehicular/robotic monitoring and control, PC to host communications etc. There is considerable current interest in vehicular location and traffic safety communications of an automatic nature which could potentially be part of FPLMTS services.

Both terrestrial and satellite systems are being studied by Task Group 8/1 since an integrated approach, based on using existing and planned fixed and mobile infrastructures where appropriate, is essential to the realization of ubiquitous personal telecommunications at a realistic cost.

The Task Group considers that it has made significant progress with its studies of FPLMTS, and the future work plan involves preparation of additional detailed recommendations on FPLMTS, some of which will be ready for presentation to the meeting of Study Group 8 in April 1992.

Spectrum Allocations for Personal Telecommunications

800/900MHz cellular allocations are presently different in the three ITU regions, and none correspond directly with the ITU spectrum recommendations for primary status in all three regions. This is not too serious for today's predominantly vehicular mobile systems, but it is not optimum for future personal telecommunications.

The ITU organizes general WARC's, dealing with all spectrum requirements for all services, every 15 to 20 years. The last one was in 1979 and there is unlikely to be another one this century. Specialized WARC's are held in the intervening periods to deal with the requirements of specific services.

WARC-MOB-87, a special WARC in late 1987 dealing only with mobile services, recommended that the next competent WARC allocate spectrum for FPLMTS and mobile satellite services.

A special WARC is to be held in the first quarter of 1992 and will, amongst other things, address allocations in the band 0.5 - 3 GHz for the land mobile, mobile satellite, direct broadcasting satellite, space research and exploration services.

Task Group 8/1 estimates that a minimum of approximately 230 MHz is required for all voice and non-voice FPLMTS Services. 60 MHz is for the Personal Station (R2) radio interface and the balance for the Mobile Station (R1) interface. These estimates include a total of 65 MHz for non-voice services. Spectrum needs for the Mobile Satellite (R3) interface and any terrestrial air-ground services have not been included as these are being considered by IWP8/14.

Fortunately the microcellular infrastructures envisaged for FPLMTS result in realistic overall spectrum needs and also the ability to provide relatively wideband wireless services.[25][28]

Conclusions

The basic structure for FPLMTS can be provided by a system architecture which evolves from present-day fixed and mobile networks toward the goal of world-wide communications mobility for people wherever they may be, at home or away, in planes, trains, cars, ships etc.

Radio system technology, VLSI, ISDN concepts, electronic switches, computer communications networks and CCITT Signalling System No. 7 all provide building blocks for this evolution.

However the necessary global focus on developing a universal personal telecommunications network will not be achieved until radio spectrum has been allocated for this purpose. The 1992 WARC provides a unique opportunity to achieve this global focus and thus ensure the necessary support for the future work of CCIR Task Group 8/1 and the many other groups involved in development of these important new wireless access standards.

It is important to remember that the customer is not interested in the technology involved, only in the cost and quality of the services provided.

The cost of a given service may vary dependent on location, even the range of services available may vary, but there is obviously a practical limit from a financial, operational and space point of view, to the number of different "personal" radio systems.

It has been suggested that in the developed countries half the adult population will have some form of personal wireless telecommunications device by the start of the next century. [16]

Common world-wide frequencies and standards for personal telecommunications should be our goal to make this new technological revolution a reality.

Acknowledgements

The support from the working members of Task Group 8/1, who have developed many of the ideas expressed in this paper, is gratefully acknowledged.

References

[1] Mallinder, B. J. T., "An Overview of the GSM System"DMR-III, 12-15 October 1988, Copenhagen, Denmark.

[2] Akerberg, D., "Properties of a TDMA Pico Cellular Office Communication System" IEEE Global Telecommunications Conference, November 28 - December 1 1988, Hollywood, Florida, USA, pp. 1343-1349.

[3] Potter, A.R., "From Cordless Through to Personal Communications" EUROCON 88, 13-17 June 1988, Stockholm, Sweden, pp. 11-13.

[4] Hattori, T., Sasaki, A., Momma, K., "A New Mobile Communication System using Autonomous Radio Link Control with Decentralized Base Stations" IEEE VTC, 1-3 June 1987, Tampa, Florida, USA, pp. 579-586.

[5] Ochsner, H., "DECT - Digital European Cordless Telecommunications" IEEE VTC, 1-3 May 1989, San Fransisco, USA, pp. 718-721.

[6] Cox, D.C., Arnold, H.W., Porter, P.T., "Universal Digital Portable Communications: A System Perspective" IEEE J S A C, Special Issue on Portable and Mobile Communications, June 1987, pp. 764-773.

[7] Gibson, R.W., "Towards a Universal Mobile Communications System" Workshop on Third Generation Wireless Information Networks, 15-16 June 1989, Rutgers University, Piscataway, New Jersey, USA.

[8] Phillips, R.O., Wright, D., "Complementarity between Terrestrial and Satellite Mobile Radio Systems" DMR-III, 12-15 September 1988, Copenhagen, Denmark.

[9] Callendar, M.H., Maclatchy, A., "Future Public Land Mobile Telecommunication Systems - A Canadian Perspective" DMR-II, 14-16 October 1986, Stockholm, Sweden, pp. 12-17.

[10] De Brito, J.S., "Personal and Mobile Communications" IEEE ICC, June 14-18 1981, Denver, Colorado, USA, pp. 57.1.1-57.1.3.

40

[11] Carpenter, P., "From Mobile to Personal Communications" DMR-III, 12-15 September 1988, Copenhagen, Denmark.

[12] Chien, E.S.K., Goodman, D.J., Russell, J.E., "Cellular Access Digital Network (CADN): Wireless Access to Networks of the Future" IEEE Communications Magazine, June 1987, pp. 22-31.

[13] Urie, A., Coutts, R., "Personal Telecommunications - The Need for Multiple Air Interfaces" IEEE VTC, 15-17 June 1988, Philadelphia, Pennsylvania, USA, pp. 52-56.

[14] Ogawa, K., Kinoshita, K., Nakajima, N., Hata, M., "Spectrum Efficient Channel Structure and Zone Structure for SCPC/FDMA Digital Mobile Radio System" DMR-III, 12-15 September 1988, Copenhagen, Denmark.

[15] Goodman, D. J., Sanjiv, N., "Technologies for Personal Communications Services" NCF, 2-4 October 1989, Chicago, Illinois, USA, pp. 882-887.

[16] Ross, M.H., "What is Personal Communications" NCF, 2-4 October 1989, Chicago, Illinois, USA, pp. 859-868.

[17] Murty, B.S., "A Phone in Every Village" Telecommunications Research Center, Department of Telecommunications, India, 1987.

[18] Canas, A. F., "The Socio-Economic Impact of Telecommunications in Costa Rica" 5th World Telecommunications Forum, October 1987, Geneva, Switzerland.

[19] Saunders, R.J., Warford, J.J., Wellenius, B.,"Telecommunications and Economic Development" A World Bank Publication.

[20] Callendar, M.H., Kenward, G.W., "Design Trends and Service Capabilities of Subscriber Land Mobile Equipment" Telecommunication Journal (ITU), Special Issue on Mobile Communications, June 1987.

[21] Maseng, T., Trandem, O., "Adaptive Digital Phase Modulation" DMR-II, 14-16 October 1986, Stockholm, Sweden, pp. 64-69.

[22] Goodman, D.J., Wei, S.X., "Factors Affecting the Bandwidth Efficiency of Packet Reservation Multiple Access" IEEE VTC, 1-3 May 1989, San Fransisco, USA, pp. 292-299.

[23] Potter, A. R., Green, E., Baran, A., Chia, S. T. S.,Steele, R., "Increasing the Capacity of a Digital Cellular Radio System by using Microcellular Techniques" International Conference on Digital Land Mobile Radio Communications, 30 June - 3 July 1987, Venice, Italy, pp. 393-402.

[24] Dornstetter, J., Verhulst, D., "Cellular Efficiency with Slow Frequency Hopping: Analysis of Digital SFH900 Mobile System" IEEE J S A C, Special Issue on Portable and Mobile Communications, June 1987.

[25] Acampora, A.S., Winters, J.H., "A Wireless Network for Wideband Indoor Communications" IEEE J S A C, Special Issue on Portable and Mobile Communications, June 1987.

[26] Callendar, M.H., "International Standards for Personal Communications" IEEE VTC, 1-3 May 1989, San Fransisco, California, USA, pp. 722-728.

[27] Sakamoto, M., Ogawa, K., "Location Probability Estimation of Service Availability in Portable Radio Telephone Systems" IEEE VTC, 15-17 June 1988, Philadelphia, Pennsylvania, USA, pp. 575-581.

[28] Callendar, M.H., "Future Public Land Mobile Telecommunication Systems - A North American Perspective" International Conference on Digital Land Mobile Radio Communications, 30 June - 3 July 1987, Venice, Italy, pp. 73-83.

WIN with OSI

Alistair Munro, BSc, PhD,
Centre for Communications Research,
Faculty of Engineering,
University of Bristol,
Queen's Building
University Walk,
Bristol, BS8 1TR
UK

Abstract

What do Wireless Information Networks and Open Systems Interconnection mean for each other? From the point of view of the OSI standardization process, looking down as many standards people would do, WIN's are another communications medium and need to be fitted into the grand scheme in due course, after suitable deliberations. From the WIN Standpoint, looking up or from the side, there is the challenge of moving a lot of data around efficiently and reliably to provide a useful service.

Computer networking on wireless media is well established whether public, corporate, as an academic experiment or for a hobby. Our experience with organizations and colleagues in this area is that the range of services and options offered by OSI standards is so large that it is difficult to identify key strategic and technical issues and their impact if implemented; more important the primary objective of open networking is undermined by the danger of incompatible choice of options and profiles.

Academic Community networking in Great Britain has been a qualified success afar as adopting and implementing standards is concerned and a comparison between it and the situation in other countries and organizations is a good indicator of the problems that can be expected as the transition to OSI gets under way. Many of the experiences and current trends apply to WIN's.

Using this as a background we describe our view of the strategic and technical relationship between WIN's and OSI and some of the important problem areas, with the aim of understanding where we are now and where we will be in twenty years time.

1. Introduction

The family of ISO standards for OSI promotes distributed information processing between dissimilar computer systems interconnected by networks of differing topologies and differing technologies. Wireless information networks (WIN's) will be used to carry OSI traffic. We believe the following factors are important in the mutal relationship between WIN and OSI:

Standards and Profiles -

> The scope of OSI standards splits (roughly) into three areas: a "filling" of communications services and protocols sandwiched between applications and media. The "filling" comes in a limited range of flavours but there are numerous options, some of which soak into the outer layers more than others and make the boundaries more or less mushy. Understanding, exploiting and using the communications standards is the key to achieving interworking with OSI.

Stability -

> OSI, in its present form, was conceived in the 1970's; but few of the standards settled down until the mid-1980's when the first implementations became available. As we enter the 1990's most of the communications "filling", several well-established networking technologies and "traditional" applications are acceptably stable. Debate and development continues under pressure from new media and better understanding of what constitutes distributed processing so change is constant.

Timescales -

> Long timescales have allowed proprietary networking, (in both the industrial and institutional sense), to become entrenched: the standardisation process tends to concentrate more on political issues, as organisations protect their investment, than on technical matters and prolongs the timescale further. The widespread, and increasing, use of proprietary standards is balanced by a convergence towards OSI but several applications have become popular that are not supported by any standard. Change, while constant, is slow as well.

Architecture -

> OSI has a well-defined, well-publicised, layered communications architecture. There are many WIN architectures; some are already aligned with the basic OSI model; others offer messaging services as well as standard interfaces; still others just provide a physical medium.

Security -

> The integrity of a network can be compromised in many ways through malice, incompetence or accident. The consequences are wide ranging if OSI applications are affected.

Performance -

> Finally, many consider that OSI standards are unnecessarily complex and that performance will be limited. It is not easy to present convincing counter-examples at this stage but these claims are not inevitably true. The demand made by typical networking applications varies widely.

The aim of this paper is to investigate how these factors will influence the relationship between WIN and OSI. The discussion is based on experiences in the academic community in the transition from de-facto or interim standards to OSI with fixed networking - we believe that the technical and strategic problems in this area will apply to WIN's as well.

2. Open Systems Interconnection - OSI

Here we discuss the concepts of OSI briefly to identify the factors that are likely to be of interest to WIN environments.

2.1. Reference Model(s) and Architectures

The OSI Reference Model, ISO 7498, parts 1-4, [1], describes the overall architecture. It includes the layered model, (Fig. 1), naming and addressing, security, and management.

The layered model provides the framework for OSI: it is important to remember that it is a *framework* and not a description of the components of conforming implementations. In particular, the layers and interfaces need not be individually accessible or identifiable in an implementation and this can have significant implications for an application's performance - it may be possible to coalesce some layers and reduce overheads. However, it *does* define a set of rules and services that OSI end open systems must obey if they claim conformance to OSI.

Recalling the sandwich analogy, the "filling" represents layers 2 (link) to 6 (presentation), and is *relatively* stable. Layers 1 (physical) and 7 (application) are under constant attack.

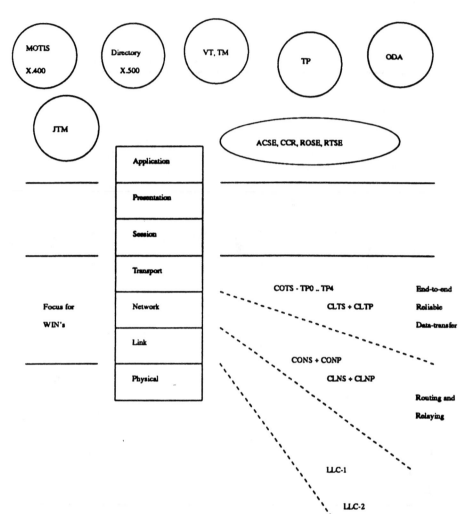

Fig. 1: OSI Communications Architecture

Connection-mode and Connection-less Services

The reference model is split into a connection-mode and a connection-less domain. In principle any layer may offer either or both to the layer above; in practice conversion at layer 5 (session) and above is not permitted. As far as current standards are concerned the arguments about the choice of service concentrate on the lower three layers.

Naming and Addressing

There are three issues: *naming*, (what systems are called); *addressing*, (where the systems are), and *routing*, (how to get to them).

The naming and addressing scheme is uniform, in an abstract sense, across all layers of the reference model. It identifies objects: these are *basic* components, (real open systems, addresses, layers, syntaxes, applications), or *local* or *dynamic* components, (selectors, associations, processes - instantiations of applications).

Objects are not specified in a form that is useful to a user and are thus made available by means of *names*.

Names are *primitive*,descriptive (complete - identifying one object or incomplete - identifying a set of objects), *generic*, (bound to a set of objects i.e. primitive or non-complete descriptive), or *specific*, (bound to one object, i.e. primitive or complete descriptive). Primitive namnes may be used as components of descriptive names. Finally, names are divided into *titles*, (naming systems, entities, or applications), and *identifiers*, (naming instantiations of entities or applications).

Entities and their invocations are the active elements in OSI; they have to be addressed as well as named. An address at a given layer in the model is used to locate entities in the layer above; these entities are identified to the layer below by *service-access-points*, (SAP's) and these live at the boundary between the layers. The SAP's themselves are located by SAP-addresses. An address at a given layer identifies a set of SAPS at that layer.

This scheme is centred round the network layer: above it all the entities reside in one system and are assigned locally defined and unique addresses to locate them. These addresses are called selectors and are carried in the control information fields of the protocols for the respective layers.

At and below the network layer communication with other systems is involved and the addressing has global significance. There is also an important extra component: the *subnetwork point of attachment*, (SNPA) and the SNPA address, which may refer to an end system, an intermediate system, or another subnetwork. As the connectivity between systems and subnetworks may be arbitrarily complex and the relationship between network SAP's and SNPA's is many-to-many.

Fig. 2 shows an illustrative example.

Security

The reference model security architecture makes wide provision for protection of data transferred between open systems and networks. Application frequently specify regimes for authenticating activities in a less communications-oriented way.

Applications

Application layer standards have varying stability. There are three families: common application service elements, application layer structure(s), and the applications themselves. The standards are listed in the references.

The common services include association-control (ACSE); commitment, concurrency and recovery, (CCR); remote-operations (ROSE); reliable-transfer (RTSE). ACSE and ROSE often provide the infrastructure for home-grown applications. CCR provides a framework for maintaining consistent relationships between applications after communications or other failures.

The application layer structure (ALS, XALS) provides rules for constructing distributed applications.

Applications include file transfer, access and management (FTAM), message transfer (MOTIS), the directory, job-transfer and manipulation (JTM), and virtual terminal, (VT, TM). We will not try to describe the full baroque splendour of these standards: many of their attributes have little or nothing to do with communication but they make heavy demands on the end systems that support them.

Toolkits to support quick implementations of applications using ROSE are widely available. This opens the way for development of new and interesting services, possibly for systems of limited capacity (eg. PCN terminals) to interface to MOTIS and the Directory via an application layer relay for instance. It also opens the way for serious interworking problems but avoids the standardisation process.

Future Directions

The reference model is a networking, or communications, model and not a distributed processing model. The latter is being developed under the name of Open Distributed Processing, (ODP, [2]), and reflects advances in object-oriented models of computing and generally better understanding of distributed processing. ODP

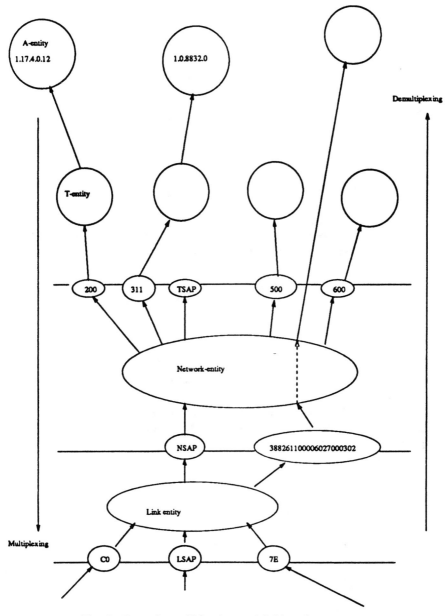

Fig. 2: Overview of Naming and Addressing

standards are still in their infancy and it is difficult to make any predictions of their final form.

2.2. Focus on the Transport and Network Services

The transport and network layers are the centre of strategic and technical discussions; the transport layer plays a pivotal role in the model. For both connection-mode, (COTS), and connection-less transmission (CLTS), its purpose is:

- to provide end-to-end transparent transfer of data between session-entities,

- to optimize the use of the network service for minimum cost, taking into account the demands of all concurrent session-entities and the overall capacity and quality of the underlying network service.

Each transport connection is associated with exactly one session-entity at any any one time; but one transport-connection may be used by more than one consective session-entity and one session-entity may use more than one consecutive transport-connection.

The relationship between transport-connections and network-connections can be one-to-many (*split*) or many-to-one (*multiplexed*), or one-to-one.

The purpose of the connection-mode network service is: to provide the means to establish maintain and terminate network connections between open systems and to exchange data between them. For both types of service, the network layer makes routing and relaying considerations invisible to the transport-layer entities.

Much of the technical debate focuses on the relative merits of a connection-mode network service, (CONS - the "O" means "oriented"), versus a connectionless one, (CLNS), and often soaks down into the corresponding debate in the link layer. Whatever conclusions are reached, it is a fact that suppliers are biassed towards CLNS, of which more below.

To permit a uniform transport service to be provided over the different kinds of network service, five classes of connection-oriented transport *protocol* have been defined. The choice of class depends on the quality of the network service and this is classified as follows:

Type A: Acceptable residual error rate; acceptable signalled error rate.

Type B: Acceptable residual error rate but unacceptable signalled error rate.

Type C: Unacceptable residual error rate.

The classes or protocol are:

TP0, (simple, type A networks)

Provides connection establishment with negotiation, data-transfer and protocol error reporting, flow-control and disconnection provided by the network service. It is often the choice to be run on a CONS.

TP1, (basic error recovery, type B networks)

Provides the functionality of class 0, together with error recovery, expedited data-transfer, consecutive transport connections on a network connection. The error recovery does not involve the service user.

TP2, (multiplexing, type A networks)

Provides transport connections with or without individual flow control; no error detection or recovery is performed. Network connection resets and disconnects cause the transport connection(s) to be terminated and the service user is informed.

TP3, (multiplexing with error recovery, type B networks)

Provides the functionality of class 2 (with explicit flow control) and recovery from failures signalled by the network layer. The service user is not involved in the recovery.

TP4, (error detection and recovery, type C networks)

This provides the functionality of class 3 with the ability to detect and recover from lost, duplicated, out-of-sequence and damaged data units. It can also use multiple network connections. This class is often used on a CLNS.

Both the network and the transport services are charged with negotiating and maintaining an agreed quality of service, (QOS). This includes residual error rate, service availability, reliability, throughput, transit and connection delays for connection-mode services. QOS for connection-less services has a wider scope as problems can propagate higher up the stack. It includes probabilities for corruption, loss or duplication, out-of-sequence and wrong delivery of data as well as measures of throughput, cost, delay and security.

2.3. Standards and Profiles

The standardisation process is long and complex and influenced as much by conflict and compromise between national standards bodies and computer manufactureres as by technical considerations. The number of options and configurations of the standards reflects this and affects, often adversely, the prospects for interworking between different and perfectly conformant implementations.

Organisations publish profiles that define the service characteristics they offer and accept, (US-GOSIP, UK-GOSIP, [3,4]) and this highlights the important role of functional standardisation.

It is always preferable to publish a set of functional characteristics of an existing standard than to embark on producing a new standard. Quite apart from timescales there are administrative problems of acceptance by ISO of the proposed standard, obtaining support for it in the international community, registering it with the various authorities.

2.4. Specification, Design and Implementation Issues

The modelling of distributed systems is one of the most vigorous areas of research in computer science. There are three main fields: processes; semantics; and logics. Major work has been done since the late 1960's, based on fundamental results obtained decades ago, (at least) and now is mature enough that languages and toolkits are available.

ISO JTC1 has adopted some of these results and defined two standards: LOTOS, [17], and ESTELLE, [18]. JTC1 calls these "formal description techniques" (FDT's) and has issued a directive that any OSI standard must be accompanied by a corresponding FD. There is a large program of work on developing the standards, interfaces to the toolkits and reasoning support systems.

Inevitably there is serious political and technical argument about which FDT should be used, whether a FD should be normative or informative, what use can it be put to. Setting this aside, we believe that these modelling methods complement current performance analysis tools and that there is an urgent need for investigating the potential for combining them.

2.4.1. Formal Description Using LOTOS

This section illustrates the concepts of LOTOS as an example of a FDT.

Processes

Computer processes can be constructed from an alphabet of actions, using a small set of operators and recursion. The rules for building processes combined with various axioms about their behaviour constitute an algebra. This together with a set of inference rules, (usually), provides a framework for proving a limited number of properties and for transforming processes from one form into another.

The descriptions of services and protocols in the standards documents can usually be converted into this form fairly easily and the behavioural part of LOTOS fulfils this role.

A simple action model is usually inadequate as processes generally manipulate data-structures and their behaviour is often controlled by the state of variables or the values exchanged between them. To be complete therefore the description languages must contain data-type constructors as well.

LOTOS has no model of time: the development of process algebras that contain a concept of time is an immature area.

Semantics

LOTOS semantics is based on a transition system, which is a set of states and actions, and a relationship between them. The construction of the transition system from the process behaviour is a long process but yields a structure that can be examined to investigate the possible interactions between processes, or exercised to achieve a kind of simulation.

At the moment a simulation must be driven manually - the operator selects the events more, or less, systematically. To avoid bias some programmable and repeatable method must be used, either probabilistic or enumerative. Both approaches appear to lead to intractable computational complexity in time and space.

Logics

For discrete systems of the type above, the most interesting questions are usually related to liveness, freedom from deadlock and can be translated into questions about occurrence or otherwise of events and sequences of events.

Posing questions in traditional propositional logic and predicate calculus is inadequate in this case as the system is not static but moving in time through a state space. Accordingly the questions must be phrased using propositions in dynamic or temporal logic.

2.5. WIN Relevance

A WIN that offers OSI services must establish its relationship with the OSI model in respect of its architecture. The level of correspondance must be identified clearly. The distinction between functional and base standardisation must be understood and applied to protocol and service design.

The transport and network layers are the focus of the relationship. The quality and characteristics of the underlying WIN will determine the choice of connection-mode or connection-less service and the naming, addressing and routing requirements. It may turn out that the WIN does not match up at all and that a non-open access to OSI services is required.

Modelling for verification is an important topic.

3. Computer Data Networking Now: WIN, Non-WIN

The use of wireless communication technologies for data transfer is well established and wide-ranging, from those technologies where the capability is designed in to those where (more-or-less) ad-hoc methods are used. The outside observer sees a mixture of standards and techniques, some regional, others international. A similar diversity characterises fixed networks.

Short-duration, limited data-length, messaging protocols, current and proposed, are particularly important in the WIN domain. The ability to transmit brief data messages is widespread and achieved by several different methods; the service may co-exist with standard interfaces.

Within the scope of this paper we consider that the relevant feature, in the context of WIN's with OSI, is the character of the data communication, that is, that the traffic is generated from the end-to-end communication arising from normal distributed processing functions of networked computer systems. This includes both those initiated by computer users as well as those that are an integral part of the computers' operating systems. Within this domain there is a mixture of international, interim, de-facto, and proprietary standards.

The international academic community is a big operator of fixed computer networks and suffers greatly from the consequences of this mixture compounded by inconsistencies between national policies and (international) research community requirements.

To illustrate this let us look at three examples in the UK, which is one of the few academic communities to have used all of these techniques on a large scale: interim protocols based on international standards (UK Coloured Books); a de-facto standard (US DoD Internet protocols); and a proprietary product (Digital Equipment Corporation's DECNet ††). The type of application that is supported is broadly similar but the communications philosophy significantly different.

3.1. Coloured Books (CB)

CB applications are used almost exclusively by the UK academic and research community.

CB protocols operate on X.25 networks and are supported by three types of network service:

-   X.25(80) and a CB-specific transport service (Yellow Book - YBTS) on packet-switched wide-area networks. YBTS is roughly equivalent to CONS, and thus *not* an transport service, and provides a distributed, source routed addressing scheme for global identification and location of end systems.

-   CONS, provided by X.25(84) on wide-area networks. NSAP addressing is used to identify end system applications globally, with DTE routing.

-   CONS with LLC-2 on local-area CSMA-CD networks, ie. base-band 10Mbit Ethernets, (not IEEE 802.3 framing on the whole). Again NSAP addressing is used to identify applications with Ethernet MAC addresses identifying SNPA's.

CB Operation

The operational style of CB applications is very much in the connection-mode mould: the lifetime of the communication is limited and, once connected, a steady stream of data flows between the applications - terminal connection is the exception although end system session management may close idle connections. There is a

---

†† DECNet is a registered trademark of the Digital Equipment Corporation.

constant overhead in traffic incurred in maintaining the CONS.

Transition to OSI for CB

There is a clear transition to OSI services for CB. This is under way at the network layer where the remaining X.25(80) networks are being superseded, although slowly, by X.25(84) and CONS.

CB applications: mail, file-transfer and job-transfer all have direct OSI equivalents so there is not transition target gap.

Protocol converters are installed where services between end systems are out of step.

3.2. Internet Protocols (IP, TCP, UDP)

The Internet is a network of networks using protocols developed from the US DoD ARPA-net. There is a large range of applications and the Internet extends to many countries, including the UK where Internet protocols are preferred in many sections of the community over CB (cf. above).

IP provides a connection-less datagram service. The higher protocols use it: the reliable stream transport protocol (TCP), the user datagram protocol (UDP), and various protocols for distributing routing information, (GGP, EGP, IGP), among others.

IP

IP corresponds to a CLNS in OSI terms; in fact the two are the same. IP operates on numerous network media and is the preferred choice for commercial products targetted at the UNIX† market. In particular it can be carried, by encapsulation in CONS, on X.25 networks.

Internet addressing is designed to support both networking and internetworking.

TCP

TCP provides a COTS over IP; it is identical to OSI COTS/TP4. It provides a reliable transport service and is used by the Internet file-transfer (FTP), the most commonly used mail protocol (SMTP), and the remote terminal service (rlogin). It is worth mentioning that one of the earliest available OSI development

---

† UNIX is a trademark of Bell Laboratories.

environments, ISODE, [21], uses TCP/IP as its basic infrastructure.

There are many TCP implementations; most can interwork but there are wide variations in timer management, congestion control and other management algorithms.

Other protocols using IP

The routing protocols use IP directly. Many other services use UDP which may be compatible with CLTS. These include above all the file-serving protocols upon which large user-communities depend.

Internet Operation and Traffic

Many Internet services have connection-mode characteristics, whether they are TCP or UDP based; others, including the file-serving and routing distributions services are connection-less.

IP networks suffer from congestion of two kinds: faulty implementation and management can cause TCP to retransmit over-eargerly; and there is a tendency for large amounts of routing information to pass around. The former case can be fixed - the algorithms are improving continuously. The latter can be avoided to some extent by adjusting the network structure - this may not be possible.

There is thus potential for continuous long-term load in an IP network. Where the IP traffic is carried in a CONS tunnel the load will appear on the X.25 network.

IP Transition to OSI

As noted above TCP + IP have a transition target of COTS/TP4 + CLNS; CLTS is a possible target for UDP. TCP based services have OSI equivalents; many UDP based ones do not. File serving is one of these and its importance cannot be underestimated.

3.3. DECNet

DECNet protocols have several similarities with the Internet protocol set. A reliable end-to-end connection is provided by the Network Services Protocol, (NSP), and this is carried in datagrams in the connection-less routing layer. Higher layer services are carried in NSP for file transfer and file access, and terminal connection. DECNet can tunnel through CONS in the same way as IP does.

There are several differences as well. Several different and self-contained protocols exist at the same level as the routing layer. These provide routing-table updates, diagnostics, cluster and terminal serving.

DECNet addressing appears relatively simple and restricted. Each node is assigned a number and is a member of an area.

DECNet operation and traffic

As with Internet protocols, some services are connection-mode, others connection-less. The routing protocols and management traffic generate a continuous load that can flood the network if timer-values are altered indiscriminately. Where DECNet tunnels over CONS it appears that network connections are kept permanently open.

Transition to OSI

One of the reasons for including DECNet here, apart from its prevalence in the scientific research community, is DEC's commitment to OSI. This means that, where possible, DECNet services will be implemented using OSI standards.

3.4. Other Issues

The numerical information in this section is drawn from informal discussions and, as yet, unpublished reports. The figures are hotly disputed in most instances.

Throughput

The throughput demanded by various services varies enormously. The most demanding are file-serving (> 64kbps), and window-managed sessions (>19.2 kbps). These rates are deemed to be the minimum acceptable. TCP/IP achieves around 700 kbps when moving images on a 10 Mbps Ethernet; this drops to around 20 kbps on a 64kbps X.25 connection.

Several experiments have been done to compare services using IP, CONS and a full-stack. The results are contentious and vary from 5 - 7% degradation as much as 90% when changing from IP to the full stack, perhaps indicating that the performance could be comparable but that some implementations could not do the job.

Traffic estimates

We quote the information that is available at the time of writing. This is more parochial than we would like - the Internet is several orders of magnitude larger! Also, it does not take account of new OSI services such as the Directory that will create a large load that cannot be assessed now.

The Joint Academic Network in the UK (JANET) is a network of several hundred hosts using X.25 on wide-area networks and CONS + LLC-2 on local-area networks. The traffic is almost all CB protocols and totals about 1000 Mb/day; about 600Mb of this is user data after allowing for X.25 overheads. The traffic increases on average by 75% per year and broke down in 1987 as:

Type	Volume (Mb/day)	Number/day
Mail	30	7500
File transfer	150	3000
Job transfer	120	1500
Terminal	300	9000
Total	600	21000

80% of mail and file-transfer traffic is local; the proportion of local terminal traffic is higher.

The European Commission is providing a uniform OSI network infrastructure for the European scientific and research community. This is called IXI and is part of the COSINE implementation phase. This is coming into service now, with the following operational characteristics:

> X.25(84) with NSAP addressing,
> 64 Kbps now; 2Mbps by 1991; 34 Mbps before 1994,
> 80% of access bandwidth available,
> end-to-end transit delay < 50 msec.
> The traffic estimates are:

> 200 Mbps average by 1992, 1 Gbps peaks
> 1 Mbps to US initially

User's Perception of Services

Where there is a choice between CB and IP services it has been observed that CB traffic is about 5% of the total. Where there is a choice between DECNet and CB DECNet carries 70% of the traffic.

The reason for this is the relative importance attached to the service provided to the customer. Both DECNet and IP are much richer in facilities than the equivalent CB implementations. (DECNet, in fact, is not visible to users at all, except in occasional cryptic error messages; nor is TCP/IP in a file-serving Unix environment). CB scores heavily on interworking capability but has limited geographical distribution; this plus point is being undermined rapidly by wider TCP/IP implementation and membership of the Internet, particularly by European research communities who might have preferred X.25 protocols. This supports the view very strongly that users choose a service because of its ease-of-use and familiarity and not because of any technological merit.

Security

Host systems connected to any of the three networks are the subject of constant attack from intruders, ("hackers"), through widely known and publicised loopholes. The networks themselves are open and it is easy to gain access; some hosts are vulnerable through bad administration and will often yield the information needed to let intruders move on to others.

The lower layers of protocol are not subverted often; but it is only matter of time before this becomes commonplace.

The adminstrative overhead is immense in maintaining the registration, accounting and auditing information that is necessary to detect and prevent security breaches. Ultimately, the responsibility rests with subscribers as it is their accounts and equipment that are used to gain access.

3.5. WIN Relevance

WIN's will become subnetworks of, or attched to, fixed networks of the types we have described, but how much of the above applies, if any?

The main difference is that the terminal nodes of the fixed networks are computers systems of various sizes but in general well configured. A proportion of the terminating equipment in a WIN network will be of this type and will be able to participate directly in OSI as end open systems; but the bulk of equipment will be hand-held units. At present most of these will suppport voice and very limited data-communication; soon there will be PCN terminals with more capabilities - as these are expected to form the largest component of the market we can expect that there will be competitive pressure to make OSI services available.

We can summarise the relevant conclusions to be drawn from the above examples as follows:

1.   Services are both connection-less and connection-mode at the transport layer but connectionless services dominate at the network layer. Current standards for second and third generation WIN's generally specify connection-mode network services and conversion will be necessary. However, a lighweight connection-less protocol could be supported by many of the messaging

protocols that are available.

2. The users' perception of the quality of the services determines the choice, not the networking technology: their choice is TCP/IP and this is a connection-less service. This reinforces the observation above in relationship to WIN's.

3. There is well-defined transition to OSI for some services but no equivalents for others. This may apply to WIN's to where services are already established.

4. Traffic is heavy and is growing rapidly. A significant proportion of this can be generated by lower layers of protocol. For the WIN technologies likely to carry the most data, traffic is heavy and growing too.

5. Mail will probably account for most of the OSI data traffic in a WIN initially.

6. Security will be a problem.

4. WIN with OSI

We have described the present state of data communications in the fixed networking world with particular attention to current services, OSI, and the transition to OSI. Now let us see how this applies to WIN and split the dicussion into services below (*Downwards*), and above, (*Upwards*), the transport layer; and the others, (*Sideways?*) that do not fit into the reference model framework.

4.1. Downwards

This is a general overview: there appears to be "no problem" for most second and third generation WIN's; there are other areas where the situation is not so clear. The division is between those where OSI or CCITT standards are already supported and the rest.

Note that in both cases, as far as the network layer is concerned, there is a structural question of identifying different wireless subnetworks and distinguishing and matching the network services where joins are made.

ISDN and CONS

Many WIN's aspire to offer ISDN services; many already support CCITT X.25. In this case, the fact that the network is wireless is not visible or really relevant to users.

The method of providing CONS through ISDN 'S' interfaces and post-1984 X.25 is already covered by OSI standards, [22]. There are several options for network-layer addressing, (cf. the Coloured Book transition above), and a compatible choice must be made to ensure interworking.

The ability to provide CONS leads naturally to support of COTS via TP0.

COTS/TP0 + CONS is an option that suits present-day packet-switching technology and narrowband ISDN. This may become less suitable as ISDN follows its planned development through various generations to broadband ISDN and ultimately IBCN with fast packet-switching using ATM protocols on a MAN infrastructure. The proposed packet loss and error rates in such networks are low enough that the error detection and recovery and flow-control functions in the network layer will become redundant. In this case a COTS/TP4 + CLNS or something more lightweight, (yet to be designed), will be required.

Others

The number of wireless networks is large and we have found it difficult to make useful suggestions given the range of access and allocation methods and the variable development of facilities for data transmission. In cases where the way higher layer protocols make use of a medium is fundamentally incompatible with the purpose of the network then access to OSI services be impossible. If this is not the case then the connection-mode or connection-less route can be taken according to the properties of the network: for example, versions of HDLC are common and can form the basis of a connection-mode link layer.

NB: An alternative approach for WIN's with built-in messaging protocols is described below.

We believe that COTS with protocol classes 0, 2, or 4 should be the target; the choice of CLNS or CONS should reflect the characteristics of the network.

4.2. Upwards

The application layer is the important factor above the transport layer as session and presentation entities do not generate traffic of their own accord.

The main generators of traffic, from large to small, initially, will be: messaging (using MOTIS (X.400), information services (eg. white-pages lookup via the Directory (X.500)), transaction-processing, and terminal traffic. All of these may use the Directory incidentally. The icing on the cake will be an increasing number of mini-services using ROSE.

4.3. Sideways?

So far we have limited the discussion to OSI services and assumed that WIN terminating equipments will be participating end OSI open systems.

Another option is to install OSI applications in the switching centres, (in auxiliary systems attached to the switches, possibly), and let these relay messages to and from subscriber equipment. This appears to be an attractive short-term solution, (lightweight protocols can be used in the WIN with much less traffic), but the long term burden on limited central capabilities will prove unsupportable. This scenario introduces an extra standardisation phase and associated delay in providing a service.

5. Conclusions

Returning to our original factors:

Standards and Profiles, Stability, Timescales -

Understanding, exploiting and using the communications standards is the key to achieving interworking with OSI. Avoid base standardisation in favour of functional standards in general but particularly when timescales are important.

The timescales for developing new standards will continue to be long. This will contribute to entrenchment of proprietary standards and pertetuate the need for transition programs.

The standards for the initial OSI traffic are stable enough to be useable. This includes the "core" services (transport and network) and application layer services such as mail and name-lookup via the Directory.

A uniform model of distributed processing will emerge and be implemented within the next twenty years.

Architecture -

WIN architectures should be oriented to the transport and network layers, (where a standard is not defined already). The choice between connection-mode and connection-less services depends on the characteristics of specific WIN's, with due consideration of the overheads.

Gateway models have a short term appeal but will be difficult to maintain and support in the long term.

Security

Security breaches will have potentially disastrous consequences as access becomes easier and applications more distributed.

Performance -

Data communication in fixed networks requires high throughput and reliability for many applications. The initial load may not include the most demanding services but traffic is growing very rapidly.

Certain types of protocol are more prone to cause congestion than others.

We have deliberately avoided mentioning any specific instances of wireless networks in this paper and concentrated on the issues that appear to be common to all when coping with OSI. Clearly some technologies are better conceived than others but, given the level of investment in wireless communications equipment, we have tried to make the OSI architecture accessible to as many as possible.

6. References

1.   ISO 7498(1984): Information processing systems - Reference Model of Open Systems Interconnection.

2.   ISO Working Draft, Basic Reference Model for Open Distributed Processing.

3.   U.S. Government Open Systems Interconnection Profile, August 1988, U.S. Federal Information Processing Standards Publication 146.

4.   U.K. Government Open Systems Interconnection Profile, available from UK GOSIP Marketing.

5.   ISO 8571 Pts 1..4, PDAD 1,2: Information Processing Systems - OSI - File Transfer Access and Management.

6.   ISO 8649, DAD 1,2, 1989, Information Processing Systems - OSI - Service Definition for the Association Control Service Element; addenda for Peer Authentication (1), and connection-less mode service (2).

7.   ISO DIS 9041, DAD 1,2,3, 1989, Information Processing Systems - OSI - Virtual Terminal Service - Basic Class.

8. ISO DIS 10169, Pts 1..3, 1990, Information Processing Systems - OSI - Distributed Transaction Processing, Model, Service and Protocol.

9. ISO DIS 9594 Pts 1..8, 1988,89,90 - Information Processing Systems - OSI - The Directory

10. ISO 9066, Pts 1,2: Text Communication - Reliable Transfer: Model and Service Definition; Protocol.

11. ISO 9072, Pts 1,2: Text Communication - Remote Operations: Model and Service Definition; Protocol.

12. ISO 10021 Pts 1..7, 1989 - Information Processing Systems - OSI Text Communication - Message Oriented Text Interchange, (MOTIS)

13. ISO 8831(1988): Information processing Systems - OSI - Job Transfer and Manipulation Concepts and Services.

14. ISO 8832(1988)+WDAD1(1989): Information processing Systems - OSI - Specification of the Protocol for Job Transfer and Manipulation.

15. ISO 9804.3+WDAD1(1989): Information processing Systems - OSI - Service Definition for the Commitment, Concurrency and Recovery Service Element.

16. ISO 9805.4+WDAD1(1989): Information processing Systems - OSI - Protocol Definition for the Commitment, Concurrency and Recovery Service Element.

17. ISO 8807(1988): Information Processing Systems - OSI - LOTOS - A formal description technique based on the temporal ordering of observational behaviour.

18. ISO 9074(1989): Information Processing Systems - OSI - ESTELLE - A formal description technique based on an extended state transition model.

19. United Kingdom Coloured Book Protocols, obtainable from Joint Network Team, Atlas Computing Laboratory, Rutherford Appleton Laboratory, Chilton, Didcot, Oxon, UK OX11 0QX.

20. Internetworking with TCP/IP - Principles, Protocols and Architecture, Doglas Comen, Prentice-Hall International, 1988

21. The ISO Development Environment: User's Manual, Vol 1..5, Marshall T. Rose The Wollongong Group, Palo Alto, California, USA, March 23, 1989.

22.	ISO 9574, PDAD 1, 1989, Provision of the OSI connection-mode Network Service by packet mode terminal equipment connected to an ISDN

Trellis Coding for Full-Response CPM

H. V. Bims [*]   J. M. Cioffi [†]

Information Systems Laboratory

Stanford University

Stanford, CA 94305

Abstract

The GSM standard for digital cellular employs Gaussian Minimum-Shift-Keying (GMSK) as a modulation format. A method for encoding the data sequence of full-response GMSK modulator is described and analyzed. This method is also general to any full-response Continuous Phase Modulation. It uses a nonlinear precoder to encode the data stream in such a way that symbol-by-symbol detection can be used to receive the data. We also extend the precoding concept so that memoryless detection can be used for any CPM modulation. The equivalent memoryless channel then permits concatenation of standard trellis codes.

Phase-trellis codes for 4-, 8-, and 16-Level Precoded-CPM modulations are described, and their coding gains are tabulated. These results describe coding gains of up to 4dB over uncoded CPM at the same information rate.

I Introduction

In digital cellular communications, bandwidth is a precious resource that must be utilized efficiently. Continuous Phase Modulation (CPM) offers the advantage of spectral efficiency, as well as resistance to channel nonlinearity and shadow fading. Emerging digital standards use CPM-class signalling modulations such as Gaussian Minimum Shift Keying (GMSK) [1], which was incorporated into the recent standard of the Conférence Européene des Administrations des Postes et des Télécommunications (CEPT) Groupe Spéciale Mobile (GSM).

Continuous Phase Modulation (CPM) is an inherently partial-response signal modulation, which implies that its phase component is affected by the memory state of the modulator. The derivation of high-performance codes for this

---

[*]Supported in part by an AT&T Cooperative Research Fellowship.

[†]This work supported in part by NSF grant number MIP 86-57266, and by the Stanford Joint Services Electronics Program (JSEP) under contract number DAAG 29-85-K-0048.

class of modulation signals is thereby complicated by what is effectively an 'inner' convolutional code. In this paper we describe a nonlinear precoder which shadows the effects of the CPM modulator inner code on the performance of the outer code.

The phase component of a CPM signal has the following form:

$$\phi(t) = 2\pi h \sum_{i=0}^{n} a_i p(t - iT) \tag{1}$$

where h is the modulation index, each a_i is a CPM data symbol, and $p(t - iT)$ is the phase pulse response of the modulation. In this paper, we assume that h is always equal to $1/M$, where M is the size of the data symbol alphabet. A data symbol is an odd integer chosen from the range, $[-(M - 1) \ldots (M - 1)]$.

The optimal receiver design for CPM signaling is a maximum-likelihood sequence detector, which matches the received signal against the exhaustive set of infinite-length CPM signals. Practical limitations require the use of the Viterbi algorithm with finite history to decode the signal. Since the signal phase is the information-carrying portion of the CPM waveform, the Viterbi algorithm must track the progression of phase in the channel to estimate the data sequence.

When a trellis code is applied to the CPM channel, the Viterbi receiver must track the modulator state as well as the concatenated trellis code state. For this reason, the Viterbi receiver for trellis-coded CPM modulations is more complex than for trellis-codes on non-ISI channels.

II   The full-response CPM precoder

For full-response CPM signalling, the phase pulse response, $p(t)$, has the form: $p(t) = 1/2$ for $t > T$, and $p(t) = 0$ for $t \leq T$. In this case, (1) becomes:

$$\phi(t) = \frac{\pi}{M} \sum_{i=0}^{n-1} a_i + 2\frac{\pi}{M} a_n p(t - nT) \quad nT < t < (n + 1)T \tag{2}$$

for $t > nT$, where the sum is computed mod 2M. As a result, there are $2M$ possible sampling phases of the modulation, and they are uniformly distributed around the phase circle. As shown in Figure 1, the CPM modulation partitions these ending phases into two cosets, which are transmitted at alternate sampling times. However, the mapping of CPM modulator inputs to phase points on the phase circle is not time-invariant. The labeling is a function of the accumulated phase in the modulator at the start of an interval.

For a full-response CPM channel, the effect of the modulator state on the labeling of symbol waveforms can be mitigated through the use of a precoder to the modulator. The precoder subtracts the accumulated phase of the modulator

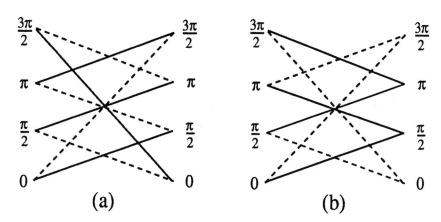

Figure 1: BMSK phase trellis diagrams. The solid line represents transmission of '1', and the dashed line represents transmission of a '0'. (a) The traditional BMSK phase trellis. (b) The precoded BMSK phase trellis.

from its input, thereby presenting an overall channel model where the labeling of ending phase states is time-invariant. Returning to the Binary MSK example, we see from Figure 1 that the labeling of trellis branches leaving the π and $3\pi/2$ states for MSK has been permuted in the precoded trellis. This is accomplished by decrementing the precoder input when the modulator is in either of those phase states. With this model, the physical trellis has been augmented such that each ending phase can be associated with a modulated data symbol.

Figure 2 shows the precoded, full-response CPM modulator. Its operation is very similar to that of a Tomlinson precoder [2] for intersymbol interference (ISI)-channels. The input sequence to the precoder, $\{\beta_n\}$, modulates a channel in which a time-invariant relation exists between the ending phase of the channel and the transmitted symbol. It is mapped into an augmented mod M sequence, $\{\alpha_n\}$, which modulates the underlying CPM channel model.

The output of the precoder, α_n, is mapped into the CPM signal set, a_n according to the following relation:

$$a_n = 2\alpha_n - (M - 1) \tag{3}$$

The phase state of the modulator is identified by x_n, since $x_n = \sum_{i=0}^{n-1} a_i$. We can therefore simplify (2) as:

$$\frac{\pi}{M} x_n + 2\frac{\pi}{M} a_n p(t - nT) \tag{4}$$

Given the fact that

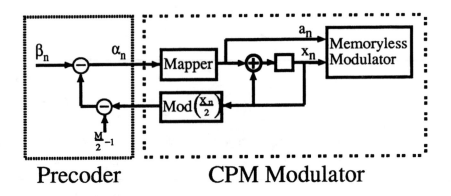

Precoder   CPM Modulator

Figure 2: Precoded full-response CPM model.

$$\{x_{n+1} = x_n + 2\alpha_n - (M-1)\} \qquad \text{mod } 2M \qquad (5)$$

for normal CPM modulation, we can derive the precoding result as follows. The desired time-invariant mapping of data symbols to phase states is described as:

$$\begin{aligned}
\{x_{n+1} &= 2\beta_n - 1\} \text{ mod } 2M & \text{for i even} \\
\{x_{n+1} &= 2\beta_n\} \text{ mod } 2M & \text{for i odd}
\end{aligned} \qquad (6)$$

Solving for the output of the precoder in terms of its inputs, we derive the relation for n even,

$$2\beta_n - 1 = x_n + 2\alpha_n - (M-1) \qquad (7)$$

$$\alpha_n = \beta_n - (\frac{x_n}{2} - (M/2 - 1)) \qquad (8)$$

and for n odd, we get

$$2\beta_n = x_n + 2\alpha_n - (M-1) \qquad (9)$$

$$\alpha_n = \beta_n - (\frac{x_n}{2} - (M/2 - \frac{1}{2})) \qquad (10)$$

$$= \beta_n - (\frac{x'_n}{2} - (M/2 - 1)) \qquad (11)$$

where $x'_n = x_n - 1$. From the phase trellis of Figure 1 and (5), we see that when n is even, then x_n is also even. Correspondingly, when n is odd, then x_n is also odd. As a result, $\frac{x_n}{2}$ for n even and $\frac{x'_n}{2}$ for n odd are always integers. Moreover, they can be computed by dropping the least significant bit of the x_n

and x'_n representations respectively. Since the two equations for even and odd time intervals have equivalent implementations, the symbol-by-symbol precoder design handles both cases.

Note also that for the Binary MSK model, the precoder degenerates into simply subtracting x_n from β_n, so the precoder effectively becomes a $\frac{1}{1+D/1-D} = 1-D$ encoder, using the most significant bit of the phase state of the modulator in the feedback path.

III Trellis code performance

For 4-level full-response CPM, the precoder feedback path is decremented mod 4 before subtraction from β_n. As a result, when a trellis code is concatenated to this precoder, the overall code will be nonlinear. This implies that even when we restrict our trellis code search to linear codes for the precoded CPM channel, we are actually searching over a class of nonlinear codes for the underlying CPM channel.

A computer program was used to compute the euclidean distance branch metrics for Ungerboeck-type trellis codes [3] applied to 4-level, 8-level, and 16-level precoded Continuous Phase Frequency Shift Keying (CPFSK), Raised Cosine (RC), and Gaussian MSK (GMSK). The trellis codes which were used are shown in Figure 3. They were chosen as examples for their simplicity in verifiying the minimum distance path by hand.

Coding gains are reported in Tables 1–3. Note that the 4-state trellis code outperforms gains reported in [4] for 4-level CPFSK. Results at higher bit rates also show significant advantage over previous work.

Since 1 bit/T is used by the codes for redundancy, the coded system actually transmits $(log_2 M) - 1$ bits/T. The coded performance is therefore compared against uncoded, $\frac{M}{2}$ ary CPM modulations. Since we are applying one-dimensional codes to the phase circle of CPM, we initially lose close to 6dB in distance by doubling the number of phase points on the circle. The coded performance overcomes this deficit to post positive coding gain advantage.

IV Conclusions

The full-response CPM precoder is a nonlinear encoder which shadows the memory effects of the CPM modulator. Its use of nonlinearity unlocks the possibility for significant coding gains when concatenated with a linear trellis code.

The trellis code-precoder combination for full-response CPM can outperform optimal linear trellis codes applied to the same modulation. An exhaustive search for optimal codes may produce codes with even better performance.

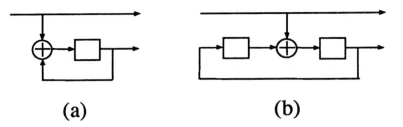

(a)   (b)

Figure 3: (a) The (3,2) trellis code used for full-response CPM modulation examples. (b) The (5,2) trellis code used for full-response CPM modulation examples.

	Rate 1/2 Trellis Code Gains on 4-level CPM				
	Uncoded d^2_{min}	(3,2) code d^2_{min}	Coding gain (dB)	(5,2) code d^2_{min}	Coding gain (dB)
CPFSK	4	6	1.76	10.546	4.21
RC	4	6	1.76	10.32	4.12
GMSK	4	6	1.76	10.53	4.2

Table 1: Coded performance gains of trellis coded-precoded 4-level CPM modulations. For GMSK, BbT = 1.

	Rate 2/3 Trellis Code Gains on 8-level CPM				
	Uncoded d^2_{min}	(3,2) code d^2_{min}	Coding gain (dB)	(5,2) code d^2_{min}	Coding gain (dB)
CPFSK	1.453	1.477	.069	3.259	3.506
RC	1.664	1.906	.589	3.772	3.554
GMSK	1.4678	1.506	.112	3.294	3.511

Table 2: Coded performance gains of trellis coded-precoded 8-level CPM modulations. For GMSK, BbT = 1.

	Rate 3/4 Trellis Code Gains on 16-level CPM				
	Uncoded d^2_{min}	(3,2) code d^2_{min}	Coding gain (dB)	(5,2) code d^2_{min}	Coding gain (dB)
CPFSK	.3988	.4	.013	.8964	3.517
RC	.4704	.5299	.052	1.06	3.528
GMSK	.4036	.4091	.059	.9075	3.519

Table 3: Coded performance gains of trellis coded-precoded 16-level CPM modulations. For GMSK, BbT = 1.

References

[1] K. Murota and K. Hirade. "GMSK modulation for digital mobile telephony". *IEEE Transactions on Comm.*, COM-29:1044–1050, July 1981.

[2] M. Tomlinson. "New Automatic Equalizers Employing Modulo Arithmetic". *Electronics Letters*, 7(3):138–139, March 1971.

[3] G. Ungerboeck. "Channel Coding with Multilevel/Phase Signals". *IEEE Transactions on Info. Theory*, IT-28(1):55–67, January 1982.

[4] B.E.Rimoldi. "Design of Coded CPFSK Modulation Systems for Bandwidth and Energy Efficiency". *IEEE Transactions on Comm.*, 37(9):897–905, September 1989.

Design Considerations for a Future Portable Multimedia Terminal

Anantha Chandrakasan, Samuel Sheng, R.W.Brodersen

Dept. of EECS,
University of California, Berkeley,
Berkeley, CA 94720

I. Introduction

Recently, the concept of "personal communications" has come to the forefront of communications research, in which individual users will have portable, private access to fixed computing facilities. The ultimate goal is to provide a personal communications system (PCS), which will move information of all kinds to and from people in all locations, through an advanced wireless network supporting a wide range of services. As a step beyond today's portable computers, a high-speed wireless link allows the advent of small, lightweight, multimedia graphics terminals, whose primary function would be to connect the user instantaneously and transparently into powerful fixed processing units and data storage. It would be capable of providing speech communication, data transfer and retrieval, computing services, and high-quality, full-motion video. Since traditional keyboard or mouse interfacing is unwieldy in a portable unit, control will be provided through speech recognition. Of course, the success of the network concept hinges upon the capability of the system to provide sufficiently high data rates. Even with the best compression schemes known today, full-motion, high-resolution digital video alone requires data rates upwards of 1Mbit/sec[12]; the terminals and network must be designed to achieve this throughput (figure 1), taking advantage of the 1+ GBit/sec capabilities of the fixed fiber-optic backbone.

We will address the hardware design issues behind such a multimedia terminal, since it is already evident that a massive amount of computing resources will be required to achieve the performance just described - the unit needs to support complex modulation schemes, sophisticated compression algorithms, and channel equalization, in addition to speech recognition and other local computing tasks. Likewise, a massive amount of power will be required to drive all of these functions. Although feature scaling of devices promises computation capability well beyond what is capable today, there are still fundamental limits to what can be achieved in terms of power. Even hypothesizing advanced packaging providing billions of transistors with multiple chips, the power requirements could easily be prohibitive for portable operation. Thus, the overriding concern lies with power consumed within the system; it is these low-power design issues which we examine, in formulating strategies for meeting the high computation requirements with acceptable power consumption levels. We begin by analyzing the sources of power consumption in the terminal, followed by a detailed analysis of the impact of power from each of the four degrees of freedom available in design: IC technology, circuit design styles, architectures, and algorithms, illustrating the extreme impact on power that can be achieved.

II. Power Consumption in a PCS Terminal

As seen in figure 2, the primary components of such a multimedia terminal are the analog RF unit/front-end, color flat-panel display, speech codec, speech recognition unit, and video decompression module. Also, the estimated operation counts (1 MOP = 1 million operations per second) and power requirements for a present-day implementation are indicated. The resulting system would require at least 20W of power, which translates to 40 pounds of batteries for approximately 10 hours of continuous operation. Clearly, more power-efficient means of accomplishing these functions need to be developed.

For the analog RF receiver/transmitter circuitry, gigahertz-band mixers, oscillators, amplifiers, and filters are of primary interest. In a recent advanced automatic gain-control amplifier capable of operating at 1 GHz[13], the power required was nearly 1/4W. Examining the circuit more closely, simulation of the core of the chip revealed that only 40 mW (a mere 20%) of this is being consumed by the multiplier itself. The remaining 80% of

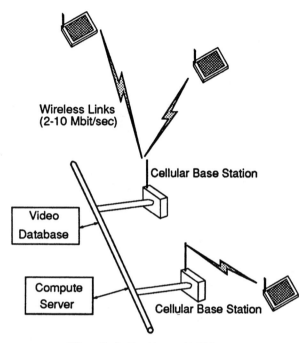

Wireless Links
(2-10 Mbit/sec)

Cellular Base Station

Video
Database

Compute
Server

Cellular Base Station

Fiber Optic Backbone (1 Gbit/sec)

Figure 1: Proposed Personal Communications System

the power was used to drive the high-capacitance pins that go off of the chip, and to buffer the input signals. Indeed, this is typical of all high-speed analog components: input and output circuitry consumes a large percentage of the power, especially when the circuitry is designed for large output voltage swings. As we shall describe below, multi-chip modules provide a feasible solution, in that a small, specialized, high-speed analog module can be "integrated" with the rest of the digital circuitry, while maintaining a low-capacitance interconnect. Given this stipulation, and the additional consideration that extreme voltage swings are not required in such a front-end, the analog RF power consumption drops to almost negligible levels compared to the rest of the system.

The display unit will take advantage of the recent advances in color, flat-panel LCD displays. Since it utilizes liquid-crystal technology, the ac-

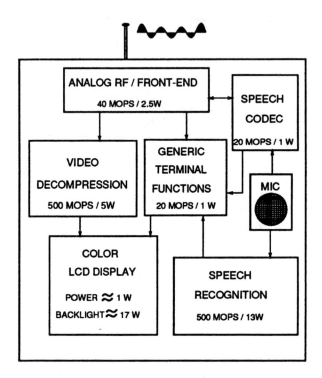

Figure 2: Multimedia Terminal

tual display consumes very little power, as the matrix is essentially an array of capacitors being slowly charged and discharged. One recent advanced display featuring a 10.4" screen size at a resolution of 640 by 480 pixels, weighed only 1.2 pounds and consumed a mere 1W of power[9, 15]. Typically, we found that the large power consumption of today's displays results from the backlighting required to enhance the image contrast - 17W for the above display. However, with improved LCD technology, this requirement will be significantly relaxed.

Lastly, the remaining functions - speech codec/recognition, general-purpose processing, data and channel encoding/decoding, and video decompression - all fall loosely under the category of CMOS digital signal processing. Given that the analog RF and display power requirements can be made relatively small, we thus focused our efforts on the sources of power dissipation and techniques for power minimization in digital VLSI circuits.

Multi Chip Modules

Implementing the system using present day technology - single chip packaging using printed circuit boards for interconnection - results in a third or more of the total power being consumed at the chip input/output (I/O), since the capacitances at the chip boundaries are typically much larger than the capacitances internal to the chip. Typical values range from a few 10's of femtofarads at the chip internal nodes, to 10's of picofarads at the chip interface attributed to pad capacitance (approximately 10pF/pin) and PCB traces (3-4 pF/in)[1, 4].

Advances in packaging technology will tremendously improve system performance by lowering power consumption, reducing system delays, and increasing packing density. The emerging multichip module (MCM) technology is one important example in that it integrates many die onto a single high-density interconnect structure, hence reducing the size of inter-chip capacitances to the same order-of-magnitude as on-chip capacitances[8, 4]. This reduces the requirements on the CMOS output drivers, and thus can reduce total power by a factor of 2 to 3. Since this style of packaging also allows chips to be placed closer together, system delays reduce and packing density increases. Extrapolating to a future 0.2 micron technology, it will thus be possible to integrate in excess of 1 billion transistors on a single 8" by 10" MCM. Hence, area constraints imposed by available silicon are no longer of great issue, allowing greater possibilities for power optimization of the various DSP algorithms.

Thus, the I/O power problem is minimized by using MCM's, leaving us to concentrate on techniques for reducing power consumption in the core of the DSP chips. We begin by presenting a brief overview of sources of power dissipation in CMOS circuits.

Sources of Power Dissipation in CMOS Circuits

There are three major sources of power dissipation in CMOS circuits: switching (dynamic), direct-path currents, and leakage currents. Usually, it is the switching component, resulting from the charging and discharging of capacitors, which dominates the total power consumption. To illustrate this, consider a simple CMOS inverter with its output loaded by capacitance C_L (figure 3). When the input switches from high to low, the output capacitance charges up through the PMOS device. In this process, $C_L \cdot V^2$ Joules of energy is drawn from the supply, half of which is stored in the capacitor while the other half is dissipated in the resistive channel of the PMOS transistor. Next, when the input switches from low to high, the charge stored in the capacitance is discharged to ground and $(1/2) \cdot C_L \cdot V^2$ Joules of energy is dissipated by the NMOS device. Thus $C_L \cdot V^2$ joules of energy are consumed per switching event or transition. The average power dissipated in this case is simply given by[10, 24]:

$$\text{Power (dynamic)} = (\text{Energy/transition}) \cdot f = C_L \cdot V^2 \cdot f$$

where f is the frequency at which the node switches. This simple analysis for power estimation holds for other complicated gates since the power dissipation is only a function of total capacitance at the output node, the voltage swing (usually the supply voltage V_{dd}), and the frequency at which the node switches.

Another important component of power dissipation results from direct-path currents[22]. Using the inverter again as an example, consider the case where the input switches from high to low. In the previous analysis, it was assumed that the NMOS device was off during the time when the output node charged up through the PMOS transistor. However, this assumption is reasonable only when the input fall time is very fast. When the input fall time is slow relative to the output rise time, both the NMOS and PMOS transistors will simultaneously be active, conducting current directly from power to ground. This short-circuit current can be kept much less than the switching component by designing for equal rise and fall times, and is given by:

$$P_{sc} = I_{avg} \cdot V_{dd}$$

Equal rise and fall times are desirable, since an extremely fast input fall time would require a large, high-power driving stage. Hence, an optimum is achieved by balancing the rise and fall times.

Finally, leakage current - mainly from substrate injection - can also be a significant source of power dissipation[23]. This component is especially significant as feature sizes scale. For a typical 1 micron device operating

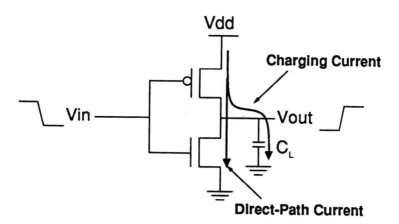

Figure 3: Power dissipation in digital CMOS circuitry

at 5V supply, the maximum substrate current is about 0.1 to 0.5 mA per micron of gate width; this value depends heavily on both technology and supply voltage. In real circuits, very few devices would operate at maximum substrate current, and we found that in practice a total substrate current of 0.1mA to 1mA is typically observed.

III. Power Minimization in Digital CMOS Circuits

We now focus on ways to reduce power consumption in the core of the digital CMOS circuitry. It is here that the richest set of design parameters are available to us: the choice of technology, logic styles, architectures, and algorithms can make a large impact on the power consumption of the multimedia terminal. These are choices which range from the low-level device characteristics, all the way up to the high-level algorithmic considerations. We shall examine each in turn, starting from the lowest level, and show how design considerations at the lower levels affect design choices at the higher ones. At all levels, though, it is clear from the above analysis minimizing power is equivalent to minimizing the expression $C \cdot V_{dd}^2 \cdot f$. Later, we will exploit the concept of parallelism heavily, and demonstrate that slower systems, using highly parallel architectures and algorithms to maintain throughput, are key elements in reducing power consumption.

Technology

When we speak of technology, we mean the use of reduced supply voltages and downscaled MOS devices to achieve our goals. By using carefully extracted SPICE device models, we extensively simulated the impact of technology on power, and the deviations from the ideal theory. In all of the analyses, we assumed that the system throughput is constant, and thus the speed of the circuitry cannot be allowed to drop without compensating for it.

Since power is proportional to the square of the supply voltage, reducing V_{dd} should be one of our primary goals. In figure 4, the effect of reducing V_{dd} on the delay of a standard 8-bit ripple-carry static adder is shown; clearly, we pay for this V_{dd} reduction in slower circuit performance, with the delay drastically increasing as V_{dd} approaches the threshold voltages of the devices. This behavior can be understood from the classical first-order theory[17, 24]: $T_{delay} = (C_L \cdot V_{dd})/I$, where (ideally) $I = k \cdot C_{ox}(V_{gs} - V_t)^2$, and $C_L = W \cdot L \cdot C_{ox}$ (typically, C_L is the gate capacitance of another MOS device). Holding the device dimensions constant for the moment, our simulated T_{delay} was inversely proportional to V_{dd}, confirming this result. Indeed, the 2.0 micron technology curve fitted the predicted theory quite well. From the results of classical scaling theory, the speed of the circuit is thus directly proportional to supply voltage, and behaves as an *inverse square*-law with respect to scaling device dimensions.

Thus, we expect that we can regain the loss in speed (since we need to maintain throughput) by scaling down the device dimensions. We also display the T_{delay} versus V_{dd} plot for a 0.6 micron technology on figure 4, demonstrating this effect. If dimensions are scaled downward by a factor of κ, Cox scales upwards by κ, and thus (holding voltage fixed), T_{delay} scales downwards as κ^2. This is plotted in figure 5 as the ideal curve, again for an 8-bit ripple-carry adder.

Unfortunately, the classical theory does not work well for submicron MOS devices[20, 3]. Two effects seriously hamper the scaled performance of circuits. The first effect is velocity saturation, where the carriers in the CMOS devices are limited in speed due to high electric fields. No longer does the current vary as V^2, but instead $I = k \cdot C_{ox}(V_{gs} - V_t)$, a significant reduction in current drive[23]. Secondly, the interconnect capacitance becomes dominant, as the interconnect ceases to scale below 1 micron due to photolithographic considerations[1]. Reexamining the 0.6 micron curve in figure 4, we found that instead of being inversely proportional to V_{dd},

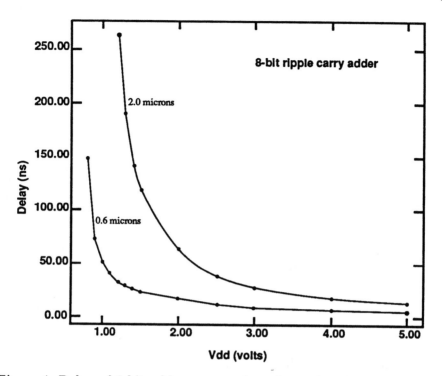

Figure 4: Delay of 8-bit adder vs. supply voltage V_{dd}, for 2.0 and 0.6 micron technologies

it is almost flat near 4 to 5V. Indeed, under the considerations of velocity saturation, the speed of the circuit becomes nearly independent of supply voltage, since the carrier velocity is only weakly dependent on electric field beyond the saturation point. One might be tempted to say this is a good effect, since supply voltage can be reduced without losing as much speed in the circuit. But, this assumes fixed device dimensions, and is only really true near high supply voltages. Once we considered the total effect of scaling the devices, as shown in figure 5, the actual delay versus technology curve went linearly with κ instead of quadratically. We did not take non-scaling interconnect capacitances into account in this analysis; as soon as interconnect and parasitic capacitances become dominant, this curve will flatten out even more. At a 5V supply voltage, the energy consumption for the 0.6 micron technology was only factor of 5 lower than that of the 2 micron technology for the same throughput - much lower than that predicted from the ideal theory.

Thus, traditional constant voltage scaling does not win as much in speed

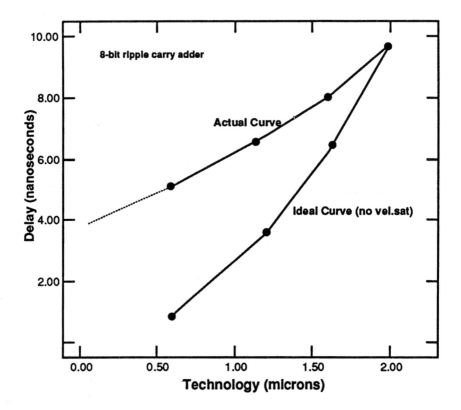

Figure 5: Delay of 8-bit adder vs. technology, ideal and actual curves

as it has in the past due to these submicron effects, and hence pure device scaling does not buy us as much power reduction as we would have hoped. However, reducing supply voltage provides significant power reduction, at a cost of lowering f. Given that the system throughput must be kept constant, compensation for this reduction in speed must be achieved by other means besides simple reliance on device scaling.

One method for maintaining a constant throughput while reducing voltage is to utilize parallelism as much as possible. For example, if the supply voltage is reduced by a factor of 2 from 5V to 2.5V, the delay increases roughly by a factor of 2 (since T_{delay} is inversely proportional to V_{dd}). In order to maintain a constant throughput, the computation can be parallelized with two parallel computation units in this case, with each one of the units operating at half the original rate. While the total capacitance has doubled with the doubling of hardware, the product $C \cdot f$ remains constant, as the units are operating at half the rate, and a net power reduction by a factor of 4 is achieved. As discussed earlier, advanced packaging technology along with feature size scaling will relax area constraints, hence making massively parallel architectures feasible.

Although parallelism can be used to compensate for such loss in speed, a certain hardware overhead is associated with this process, in that control circuitry has been added and parallel input/output capability is required. At some point, this added overhead dominates any further gains in power consumption from further voltage reduction. This leads to the existence of an "optimum voltage," such that the power dissipation is increased when the supply is reduced further. To determine this optimum voltage, the normalized power as a function of operating voltage is plotted for a given fixed throughput rate (figure 6) with a certain overhead incorporated. This voltage occurs around 1.5V for the 2 micron CMOS process. Figure 6 also shows the normalized power plotted versus operating voltage for the scaled 0.6 micron technology, which has its optimum near 1V.

Circuit Design Styles

There are a number of options available in choosing the basic circuit approach for implementing various logic and arithmetic functions. Primarily, choices between static versus dynamic, passgate logic versus traditional CMOS, and adjustment of threshold voltage are all open to the system designer. Again, to evaluate the various tradeoffs involved, we extensively simulated (using extracted circuit layouts and the SPICE simulation program) the energy and delay performance of several representative logic styles, using an 8-bit adder as a reference. As a function of supply voltage, we have examined the following logic families: DCVSL (differential cascade-voltage switch logic)[11, 24], traditional static CMOS[10, 2], an optimized static style, CPL (complementary passgate logic)[24, 18], and CPL with modified threshold voltages[25]. The results plotted on logarithmic axes in figure 7 , demonstrates the wide range of performance which can be achieved.

When considering static versus dynamic designs, it is not immediately evident which one is better. One interesting phenomenon with static designs is that they "glitch," i.e. there are spurious transitions due to propagation delay from one stage to the next. An example of this is a static 8-bit adder, with all inputs going from zero to one. The sum should be zero for all bits; unfortunately, a "one" will appear briefly at all of the outputs, since the carry takes a significant amount of time to propagate from the first bit to the last. These spurious ones all represent power being drawn

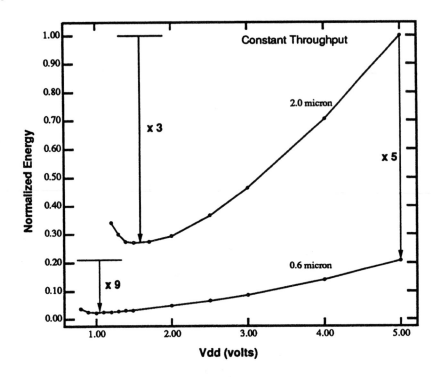

Figure 6: Optimum V_{dd}, for constant throughput and increasing parallelism

from the supply; thus, the power consumed in static circuits ends up being much higher than one expects. Also, static designs are victim to direct-path currents. Dynamic logic, on the other hand, does not have either of these problems, as the nodes are precharged on one clock cycle and evaluated on the next; thus, no direct path from supply to ground is ever allowed. Glitching is no longer an issue, since dynamic logic intrinsically requires that any node only undergo a maximum of one transition per clock cycle.

There are problems with dynamic designs, though. The power required by the drivers for the precharge signal could be quite significant; also, since nodes that evaluate to "zero" have to be pulled "high" during the precharge phase, one can imagine situations in which most nodes are constantly switching and dynamic logic could end up requiring significantly more power than static logic. We found this to be true of DCVSL, whose performance was around 2-3 times worse than all of the other families. However, in other dynamic families, it turns out to be possible to minimize the number and size of devices, to the point that a dynamic family actually has much better performance than a static one. The simulated CPL logic

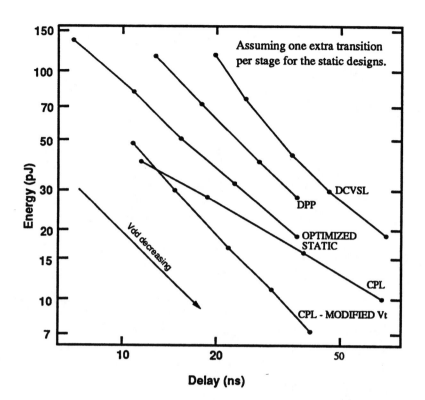

Figure 7: Energy consumed vs. delay, varying V_{dd} as a parameter

family, which used a dynamic structure, is a clear example of this.

CPL turns out to be attractive for several reasons. First, it is possible to perform an exclusive-OR operation using only two pass-transistors, instead of the multiple-transistor gates required by other logic families[24, 18]. As the XOR is the key to most arithmetic functions, it permits adders and multipliers to be created using a minimal number of devices. Likewise, multiplexers, registers, and other key building blocks are simpler in CPL. This translates to fewer nodes and less capacitance that need to charge and discharge each cycle. Correspondingly, the precharge devices can be much smaller (compared to another dynamic family like DCVSL), and thus overall circuit power consumption is noticeably reduced. CPL becomes even more attractive when the threshold voltages of the pass transistors are reduced to zero volts, thus increasing the current drive of the pass devices (and hence the speed). In other logic styles, reduction of device threshold voltages invariably results in a loss of noise margin, with a gain of current drive and speed. However, with CPL, the unique situation occurs where there are devices that only need to pass current (and not sustain a particular logic voltage level), so noise margin is not an issue here. Hence, by selectively altering the threshold voltages of the pass devices during processing, improved speed can be achieved at no extra cost in power or circuit performance.

Thus, a factor of 3 to 5x spread in power is possible as the choice of logic styles is varied. The choice of dynamic CPL with modified threshold promises to provide higher speed circuits at a lower power consumption level than the typical static or "poorer" dynamic styles used today, implementing the required logic functions using smaller, and fewer, MOS transistors.

Architectures

In the preceding analysis, we already discussed the use of parallelism in the system to permit reduced supply voltages, while maintaining throughput. Pipelining, time-shared structures (eg, general-purpose microprocessors), parallel processing by hardware replication, and bitserial techniques are all options open to the designer at the architectural level.

As a starting point, we examined a simple datapath, intended for an Euclidean metric calculation. As shown in figure 8a, inputs A and B are added, the result squared, and then latched at the output. The system throughput was given by the latch, which yielded new data every T seconds. Applying the above parallelism techniques to this simple case, we examined the effects on the power consumption of each.

With a time-shared datapath (figure 8b), we created a simple arithmetic unit which was capable of performing a squaring operation and an addition. One of the inputs was then multiplexed, and the result from the output accumulator latch was fed back to the input of the ALU. To coordinate the transfer of data, a control module was added to the circuit. Since we now had essentially half the hardware, the total capacitance was reduced by a factor of 2, minus approximately ten percent for the additional control circuitry. Now, for constant throughput, we were required to double the clocking frequency of the ALU operation in order to perform both the addition and the squaring in T seconds. However, the frequency of operation for the components only increases slightly to account for overhead circuitry. Thus, substituting $2f$, $(1.1 \cdot C/2)$, and $1.2V_{dd}$ into the power expression, net power went up by a factor of 1.6 over the original datapath. The immediate conclusion was that time-shared architectures are *not* good for reducing power. In essence, what time-sharing hardware targets is reduction in area consumption, at the expense of higher frequency. This makes general-purpose microprocessors extremely unattractive, since they typically contain only a single time-shared computation unit. Lastly, we should make some mention of bitserial structures, which are the most extreme case of time-shared architectures[6]. They occupy extremely little area, at the expense of tremendously high circuit speeds - again, not attractive for low-power applications.

With parallel hardware replication, the exact opposite of the above was seen. We now split the input data into two identical adder- squarer datapaths, as shown in figure 8c, and then multiplexed the two at the output. The hardware was doubled, doubling the total capacitance; however, the circuit ran at half the speed, since each datapath was now allowed $2T$ seconds to perform its function, and thus we could approximately halve the supply voltage. Thus, examining the power expression, the increase in capacitance canceled the decrease in frequency, while the lowered supply voltage yielded a net reduction in power by a factor 4.

Another approach was to apply pipelining to the architecture, as shown in figure 8d. In the original datapath, both the adder and the squarer had to complete their respective functions in $T/2$ seconds in order for the output to be latched every T seconds. With the addition of a pipeline latch, each operand now had a full T seconds to complete, thus permitting each element of the pipeline to operate at half the speed while maintaining total throughput; note that components still switch once every T seconds, however, yielding no reduction in frequency. There was only a slight increase in hardware (approximately ten percent) due to the latches. Thus, capacitance increased by 1.1, and the supply was halved, yielding a total reduction in power by approximately a factor of 3.6.

Both pipelining and parallel hardware represent elegant examples of the direct tradeoff between silicon area and power consumption; with advanced packaging, area is no longer of great concern, whereas power is. In today's (nonportable) systems, the opposite is true: area is of significant concern, whereas power is not. Circuits now need to operate as slowly as possible, with as low supply voltages as possible. Hardware duplication and pipelining are architectural techniques by which we can achieve this.

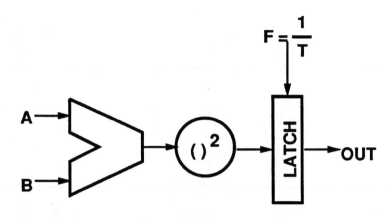

Figure 8a: Simple Datapath: Power=$C \cdot V^2 \cdot f$

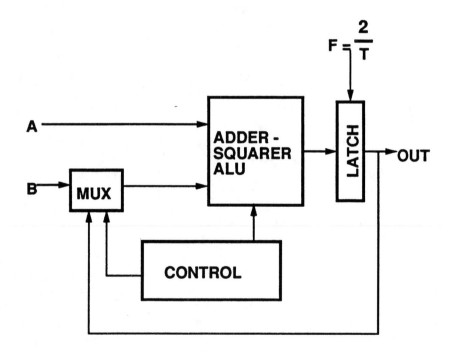

Figure 8b: Time-Shared Datapath: V_{up} is the voltage required to meet the delay constraints. Power is 1.6 times higher than the simple case.

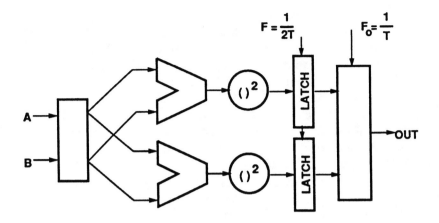

Figure 8c: Parallel Datapaths: Power is 4 times lower.

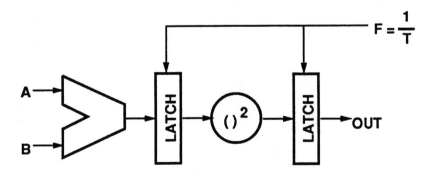

Figure 8d: Pipelined Datapath: Power is 3.6 times lower.

Algorithms

Extending our previous arguments, at the algorithmic level we can optimize in two ways: development of massively parallel implementations, and reduction in the number of operations required. With heavy parallelism, the algorithms can map directly into parallel architectures, with all of the commensurate gains. With reduced operation counts, we require less hardware and switching events, and thus lower power. Unfortunately, these two goals may be in conflict with one another, ie. optimizing operation counts may result in a highly nonparallel algorithm, or vice-versa. For example, the discrete Fourier transform can be computed by either the brute-force summation, or by the classical FFT. As it turns out, the structure of the FFT does not permit heavy parallelization, as each intermediate stages of the calculation depend on all of the previous results; however, we have significantly reduced the number of operations. Conversely, the brute-force DFT requires large numbers of operations, but is easily parallelizable.

A minimized operation, highly parallel algorithm is the most desirable, but may be unfeasible. Given that these two optimization strategies may very well be in conflict with one another, which is the "better" of the two? The obvious choice is to minimize the operation count. Even if the algorithm cannot easily be parallelized, pipelining can still be used in many cases, and in fact, pipelining is even more preferable than pure parallelism. Thus, as for minimizing complexity and increasing speed, minimizing the number of additions, multiplications, and memory accesses in the algorithm are key in reducing power consumption.

Case Study:
The Discrete Cosine Transform

Our envisioned multi-media personal communications system will provide full-motion video capabilities. In order to support full-motion video, sophisticated compression algorithms are needed to reduce the transmission bandwidth from 188 Mbits/s (for images defined over a 512 by 512 lattice assuming 24 bits per pixel and 30 frames per second) to approximately 1Mbits/sec. Transform based image compression has been proven to be an excellent method for image data compression[5]. In a typical transform coding scheme, the image is divided into small rectangular blocks (subimages), and a two dimensional transform is applied to each block. The transform represents most of the energy in the image (or subimage) by a few coefficients; by utilizing only these coefficients in transmitting and reconstructing the image, compression is achieved at the expense of slight image degradation.

Thus, the most efficient transform is one with the maximum "energy packing" capability. Of all the linear transforms, the Karhunen-Loeve transform (KLT) has the best energy compaction properties, and it turns out that the discrete cosine transform (DCT) is a good approximation of the KLT[12, 14]. This, combined with its ease of implementation, makes the DCT a very attractive and frequently used compression algorithm.

The choice of algorithm will be the most leveraged decision in meeting the power constraints. Since minimizing switching events is very crucial in minimizing power dissipation for a CMOS technology, it is important to use algorithms that require the minimal number of operations/sec while meeting design constraints. For example, there is a wide selection of algorithms for realizing an 8 by 8 DCT (figure 9), with the computation complexity varying by 15:1 from the best case to the worst case [7, 5].

IV. Conclusions

Our ultimate goal is to provide a personal communications system capable of simultaneously supporting a variety of user applications, including speech, database access, computation, full-motion video, and graphics. The centerpiece of such a system lies in the development of a small, portable terminal unit, capable of performing all of these functions. In making portability a requirement, the total allowed power consumption in the terminal is thus sharply constrained. After analyzing the power requirements of the various functions, the analog RF, the display, and the DSP functions, we have concluded that minimizing the power in the core of the CMOS computation circuitry is the key to achieving a low-power design. Techniques for minimizing power using the four degrees of freedom available to us - IC technology, circuit design style, architectures and algorithms - have been described.

At the IC technology level, progress in scaling of MOS feature sizes will indeed make possible computation far in excess of that achievable today, but there are still fundamental limits to what can be achieved. In particular, velocity saturation effects and non-ideal scaling of interconnect capacitances will limit the gain achievable from device scaling to a factor of less than 5 (going from today's 1 micron technology down to 0.2 micron).

Likewise, advances in packaging technology (such as multi-chip modules) will tremendously improve system performance by reducing power dissipation in input/output circuitry, lowering system delays, and increasing packing density. We thus projected that area can be traded off for power, since an excess of 1 billion transistors can be packaged in a single MCM. This is crucial, as the increased use parallelism requires large increases in the number of transistors needed.

There are a number of options in choosing the basic circuit approach for implementing logic and computation circuitry. Choices such as static versus dynamic, passgate versus conventional gate, and threshold voltage alteration have been extensively simulated and compared to determine what kinds of advantages can be achieved here. A dynamic passgate family, using adjusted threshold voltages was found to be extremely promising, and a factor of 5 difference in power was displayed from the worst logic family to the best.

Highly parallel architectures, either using pipelining or hardware replication, was demonstrated to be effective in maintaining system throughput, compensating for slower device speeds that resulted from reduced supply voltages. Time-shared datapaths, found in general-purpose microprocessors and bitserial implementations, were found to be unsuitable for low-power applications, as they actually require higher device speeds. Using a combination of highly parallel architectures and reduced supply voltages a factor of ten power reduction is clearly possible.

Lastly, the choice of algorithm is thus far the most highly leveraged decision in meeting the power constraints. The ability for an algorithm to be parallelized and pipelined will be critical, and the basic complexity of the computation must be highly optimized.

Thus, by utilizing a combination of the above techniques, power reductions to the level that portability is feasible can easily be envisioned. Sheer reliance on device scaling is not possible, since it no longer gains as much as it once did. Instead, the careful use of parallelism and reduced system voltages can achieve precisely the same ends.

Algorithm for 8x8 DCT	Multiplies	Additions
Brute Force	4096	4096
Row-Col DCT	1024	1024
Row-Col FFT	512	1024
Chen's Algorithm	256	416
Lee's Algorithm	192	464
Feig's Algorithm (scaled DCT)	65	493

Figure 9: Comparison of Discrete Cosine Transform Algorithms

Acknowledgements

This project was sponsored by the Defense Advanced Research Projects Agency, and monitored by the U.S. Department of Justice (Federal Bureau of Investigation), contract number J-FBI-90-073. The authors wish to thank Prof. P. Ko, A. Stoelze, A. Burstein, and others at UC Berkeley who have contributed to this project, and also wish to acknowledge the kind support of the Fannie and John Hertz Foundation.

References

[1] H.B. Bakoglu, *Circuits, Interconnections, and Packaging for VLSI*, Addison-Wesley, Menlo Park, CA, 1990.

[2] R.W. Brodersen et. al., "LagerIV Cell Library Documentation", Electronics Research Laboratory, University of California, Berkeley, June 23, 1988.

[3] A. Bellaquar, et al., "Scaling of Digital BiCMOS Circuits", *IEEE Journal of Solid-State Circuits*, Vol. SC-25, August, 1990, pp. 932-941.

[4] D. Benson, Y. Bobra, B. McWilliams, et al., "Silicon Multichip Modules", *Hot Chips Symposium III*, Santa Clara, CA, August 1990.

[5] R.J. Clarke, *Transform Coding of Images*, Academic Press, NY, 1985.

[6] P.B. Denyer and D. Renshaw, *VLSI Signal Processing: A Bit-Serial Approach*, Addison-Wesley, MA, 1985.

[7] E. Feig, "On the multiplicative complexity of the Discrete Cosine transform", *1990 SPIE/SPSE symposium of Electronic Imaging Science and Technology*, Santa Clara, CA, February, 1990.

[8] G. Geschwind and R.M. Clary, "Multichip Modules- An Overview", *Expo SMT/HiDEP 1990 Technical Proceedings*, San Jose, CA, 1990.

[9] Hitachi America, Ltd., Electron Tube Division, *Hitachi Color TFT LCD Module*.

[10] D.A. Hodges and H.G. Jackson, *Analysis and Design of Digital Integrated Circuits*, McGraw-Hill, NY, 1988.

[11] G.M. Jacobs, *Self-Timed Integrated Circuits for Digital Signal Processing*, Ph.D. Dissertation, University of California, Berkeley, 1989.

[12] A.K. Jain, *Fundamentals of Digital Image Processing*, Prentice Hall, NJ, 1989.

[13] R.P. Jindal, "Gigahertz-Band High-Gain Low-Noise AGC Amplifiers in Fine-Line NMOS", *IEEE Journal of Solid-State Circuits*, Vol. SC-22, August, 1987, pp. 512-520.

[14] J.S. Lim, *Two-Dimensional Signal and Image Processing*, Prentice Hall, NJ, 1990.

[15] P. Mannion, "Flat-Panel Onslaught Threatens CRT Stronghold", *Electronic Products*, Vol.33, No.4, September, 1990, pp. 29-34.

[16] A. Mukherjee, *Introduction to nMOS and CMOS VLSI Systems Design*, Prentice Hall, NJ, 1986.

[17] R.S. Muller and T.I. Kamins, *Device Electronics for Integrated Circuits*, John Wiley & Sons, NY, 1986.

[18] A. Rothermel, B. Hostica, et al., "Realization of Transmission-Gate Conditional Sum Adders with Low Latency Time", *IEEE Journal of Solid-State Circuits*, Vol. SC-24, June, 1989, pp. 558-561.

[19] P.A. Ruetz and R.W. Brodersen, "Architectures and design techniques for real-time image-processing IC's", *IEEE Journal of Solid-State Circuits*, Vol. SC-22, April, 1987, pp. 233-250.

[20] E. Sano, T. Takahara, and A. Iwata, "Performance Limits of Mixed Analog/Digital Circuits with Scaled MOSFET's", *IEEE Journal of Solid-State Circuits*, Vol. SC-23, August, 1988, pp. 942-948.

[21] P. Solomon, "A Comparison of Semiconductor Devices for High-Speed Logic", *Proceedings of the IEEE*, Vol. 70, No. 5, May, 1982, pp. 489-510.

[22] H.J.M. Veendrick, "Short-Circuit Dissipation of Static CMOS Circuitry and Its Impact on the Design of Buffer Circuits", *IEEE Journal of Solid-State Circuits*, Vol. SC-19, August, 1984, pp. 468-473

[23] R.K. Watts (ed.), *Submicron Integrated Circuits*, John Wiley & Sons, NY, 1989.

[24] N. Weste and K. Eshragian, *Principles of CMOS VLSI Design: A Systems Perspective*, Addison-Wesley, MA, 1988

[25] K. Yano, et al., "A 3.8ns CMOS 16x16 Multiplier Using Complementary Pass Transistor Logic", *1989 Custom Integrated Circuits Conference*

DYNAMIC RESOURCE ACQUISITION:
Distributed Carrier Allocation for TDMA Cellular Systems

Sanjiv Nanda
AT&T Bell Laboratories
Room 3M-317, Crawfords Corner Road
Holmdel NJ 07733

and
David J. Goodman
WINLAB, Rutgers University
Box 909, Piscataway NJ 08855

Abstract

Third generation wireless information networks will serve a large user population by means of a dense grid of microcells. Network control will be distributed among many dispersed processors, rather than concentrated at the mobile switching center. Teletraffic in microcells will be characterized by large variations in call arrival rates, with geographically localized peaks lasting several minutes. In this paper, we propose a simple model for microcell traffic, and consider the problem of dynamically moving the limited radio spectrum among the microcells, to accommodate the large traffic variations. We consider the allocation of carriers to microcells, where each carrier is capable of carrying several conversations (voice channels) simultaneously. Since our basic unit of allocation is a carrier, our channel allocation scheme does not react to variations in channel quality. Instead, in response to traffic variations, base stations make channel acquisition decisions using a cost/reward metric based on the carrier usage in neighboring cells. For a moderately loaded system (average traffic 0.5 Erlangs per channel) with time varying call arrival rates (peak rate ten times the average), the dynamic scheme achieves less than 2% blocking on average, compared to 10% blocking with fixed allocation. The reduction in call blocking in highly loaded cells (during traffic peaks) is even more dramatic.

1. Introduction

To deliver advanced information services to the general population, third generation wireless information networks will have to overcome many obstacles that impede the growth and enhancement of present cordless and cellular systems. Operating within a limited spectrum allocation, they will achieve high user density through dense grids of microcells (200 m spacing outdoors) and picocells (20 m spacing indoors) [1,2]. Distributed control systems will manage the large networks of tens of thousands of cells with millions of mobile users. WINLAB is currently studying packet transmission and switching techniques that will facilitate distributed network control [1]. With each packet carrying routing information, thousands of distributed processors will route the packets to their destinations. Much of the mobility management function of the mobile switching center will thus be transferred to an ensemble of these low-complexity processors. Without this decentralization, the mobile switching center would have to expand by many orders of magnitude in processing capability. Decentralized radio access technologies such as packet reservation multiple access[3,4] and code division multiple access[5] will operate in harmony with distributed packet switches.

A dense network of small cells will exhibit extreme temporal and geographical variations in traffic density. To make efficient use of limited spectrum, networks will have to rearrange transmission resources dynamically to meet rapidly changing demand for communication channels. To characterize the variations in user density, this paper introduces a simple teletraffic model for mobile services. It also explores a technique for dynamically assigning to cells, groups of transmission channels in response to changing traffic patterns. The technique is applicable to hybrid frequency-division/time-division radio access schemes. These hybrid systems consist of several carrier frequencies, each capable of supporting several simultaneous conversations (channels). We consider a carrier with several channels as our basic unit of resource allocation between base stations. In our approach, cells with busy carriers, nearing full capacity, acquire new carriers while cells with idle capacity release assigned carriers. The specific carriers acquired and released are selected on the basis of usage patterns in surrounding cells. In keeping with our distributed network control paradigm, the base stations, rather than a central controller, make carrier acquisition and release decisions.

Beck and Panzer [6], provide an excellent review of the existing work on this problem and suggest a unified setting for considering Dynamic Channel Allocation (DCA) algorithms. They suggest that intra-cell handover is a requirement for DCA to adapt effectively to variations in traffic *and* interference. Using intra-cell handovers, existing calls in a cell may be moved to new channels in the same cell, as necessary, to maximize channel reuse. If intra-cell handovers are not permitted, DCA schemes end up in inefficient reuse patterns, dictated by the specific pattern of call arrivals, call completions and inter-cell handovers. When the system load increases, the adaptive schemes with inefficient reuse have worse performance than a fixed channel allocation scheme that is designed with optimal channel reuse. When intra-cell handoffs are permitted, complete reassignment of calls to channels may be done system-wide, as often as necessary, so that an optimal assignment is always achieved. This is known as the Maximum Packing strategy. Since even a fixed optimal assignment is computationally intensive, maximum packing is clearly impractical for dynamic channel allocation.

Beck and Panzer also describe a dynamic, distributed algorithm called DYNINF, that utilizes periodic local measurements of carrier-to-interference ratio (C/I) to make decisions about channel assignments. Since base stations make autonomous decisions about channel acquisition and reassignment of calls, temporal and spatial variations in interference necessitate rapid intra-cell handovers and reassignments. The authors use hysteresis in their algorithm to reduce the number of system configuration changes. Because channel assignments are derived from local interference measurements, this algorithm does not require frequency planning. For large cellular systems this is a very attractive feature. Another algorithm that does not require frequency planning was proposed by Furuya and Akaiwa [7] and is called Channel Segregation. Here a preferable channel reuse pattern develops by an evolutionary learning process.

2. Resource Acquisition Based on Traffic Variations

In accord with the Cellular Packet Switch architecture, we describe a distributed resource allocation procedure in which base stations make carrier *acquisition* decisions autonomously. In this respect our work most closely agrees with [6]. There are some important differences. Since our basic unit of resource is a carrier, our Dynamic Resource Acquisition (DRA) technique assigns channels in groups, rather than individually. Thus it is designed to work with the PRMA protocol and other time-division schemes in which the resource to be allocated is a carrier, rather than an individual time slot.

A second contribution of this paper is a preliminary model for cellular traffic that characterizes large rapid traffic variations in individual cells. Although measurements of mobile traffic in microcells are necessary to validate such a model, these are unlikely to be available before large scale microcellular systems are established. Because measurements obtained in large cell systems of today reflect the aggregate traffic from potential microcells, it is difficult to predict microcell traffic with any degree of confidence. Moreover, today's calling patterns derive from present tariffs and the limited (voice-only) services available. What is needed is a robust model for microcellular traffic that is sufficiently flexible to characterize the measurements of tomorrow. Our preliminary cellular traffic model, described in Section 3, describes traffic details including call arrivals, inter-cell handovers, and call completions; it completely ignores propagation conditions such as signal and interference variations.

Recent work on DCA has led to the following conclusions [6-9]:
1. Fixed assignment schemes (with optimal reuse) perform better than DCA schemes (where the reuse pattern settles to a sub-optimal configuration), when the system is highly loaded in every cell.
2. DCA with intra-cell handovers is necessary to combat the unpredictable propagation and signal and interference variations in microcells.
An erroneous conclusion is sometimes drawn from the above results, as follows. Since the response of DCA schemes to high traffic is inferior to fixed allocation schemes, DCA is required and useful only as a response to signal and interference variations in microcellular networks.

In fact, the design of a dense grid of microcells ensures that at any given time all cells cannot be heavily loaded. The precise characteristic of microcells is likely to be that the time-averaged traffic in any cell is a small fraction of the short-duration peak traffic. DCA schemes are clearly required to combat these large dynamic traffic variations. The scenario where all cells are heavily loaded, and FCA outperforms DCA, will never occur in microcellular networks that are designed to achieve a small call blocking probability. In practical scenarios, DCA schemes will respond efficiently to large traffic variations in microcells and provide dramatically better performance.

Our approach to DRA begins with a fixed assignment of carriers to cells on the basis of long-term average traffic. This fixed assignment, which is applied to a fraction of the available carriers, may be based on graph coloring [10] or use an evolutionary approach [7]. The new DRA algorithm assigns the remaining carriers to cells with heavy traffic. Therefore DRA is a hybrid scheme, with both fixed and dynamically allocated carriers.

The proposed DRA scheme requires intra-cell handovers, but not in response to signal-to-interference variations. Instead, calls in any cell are reassigned to TDMA (or PRMA) channels to ensure that the carriers are fully utilized. In every cell calls are reassigned so that the existing calls are accommodated with the fewest number of carriers. A high volume of intra-cell handovers poses a serious control issue in circuit-switched systems. PRMA [3], or other systems with decentralized handovers, can achieve intra-cell handovers with minimal base station intervention and no interaction with the central switch [1]. In PRMA, base stations stimulate intra-cell handover by mobiles, by simply indicating that all slots on an unneeded carrier are reserved. Terminals then contend for new reservations on other carriers. To provide an uninterrupted reservation for talkspurt transmission, PRMA delays intra-cell handovers until the end of a talkspurt. Since the average talkspurt duration is of the order of one second, it appears that this delay will not lead to any substantial degradation in the DRA algorithm performance (assuming that we are using DRA to combat traffic variations that last several minutes). However, this needs careful study. Finally, PRMA also has the advantage of a "soft" capacity, so that delays in DRA decisions lead to marginal degradation of transmission quality rather than additional dropped or blocked calls.

While the previous paragraph indicates how the proposed algorithm meshes well with the PRMA protocol [3], we will consider a simplified and more general setting in this paper. Thus we will consider a TDMA system (with fixed capacity carriers) and assume that instantaneous intra-cell handovers are possible. Furthermore, we assume that each cell has complete carrier use information about neighbors up to twice the reuse distance, and that this information is available correctly and instantaneously. The instantaneous assumptions above may be relaxed provided the time scale for carrier reassignments (minutes) is much larger than the time scale for completing intra-cell handovers (seconds) and the time scale for communicating carrier usage information over the cellular infrastructure (fraction of a second).

In Section 4 we describe in detail the DRA algorithm and in Section 5, we apply it to a simulated one-dimensional system of microcells. This could be a model of a highway communication system. The algorithm proposed extends to two or more dimensions in a natural way. In practical resource acquisition, the main difference between a one-dimensional model and a two- or three-dimensional model is that at each cell in a higher dimensional system there are more neighboring cells (within a radius of twice the re-use distance) to consider. This translates to higher complexity and storage requirements.

In Section 5, we provide preliminary simulation results that indicate the effectiveness of this scheme, given our traffic model. These simulations are performed on the highway microcell system with a simplified version of the proposed DRA algorithm that only utilizes the carrier usage information up to one reuse distance. Due to its similarity with the Nearest Neighbor algorithm of Cox and Reudink [11], we refer to this simplified algorithm as DRA-NN. Especially in two or three dimensions DRA-NN has much lower storage requirements than DRA.

3.   Microcell Traffic Model

In this section we describe an event driven simulation model for call traffic in a microcellular system. The following are the key features of this model.

1.   We expect microcellular traffic to be non-uniform and have dynamic variations that are unpredictable. In fact, that is the motivation for our DCA algorithm. Our model provides dynamic spatial and temporal traffic variations.

2. Our model has a small number of parameters. Thus important features of the problem and the algorithm performance can be identified, and thoroughly studied by varying a few parameters. In addition, few statistical measurements will be required to validate the model and to measure these parameters. The model is

intended to be general and capable of describing a variety of scenarios. Thus traffic in highway microcell systems (the example considered in this paper), indoor, or city microcells can be generated using the same model and varying the parameters.

3. Although model validation is not possible until microcellular systems are deployed, the model provides a simulation scenario to exercise our DRA algorithm; it should also be useful for simulation studies of other DCA schemes, as well as other aspects of microcellular systems.

The traffic model only considers call arrivals, call completions and cell-boundary crossings of mobiles during calls. We refer to the latter event as a call crossover. If a voice channel is available in the cell to which the crossover occurs, we have a successful inter-cell handover. If not, we have a dropped call. Similarly, if a voice channel is not available in a cell at a call arrival event, we have a blocked call. Each arrival, completion or crossover in the system (any cell) is an event that determines the future course of the simulation. To simplify the situation, we assume that all inter-arrival times are exponentially distributed. This implies that we may simulate only these discrete events. The state of each cell is the number of calls in progress at the cell and the carrier assignment. The system state consists of the state vector of individual cell states. We assume that blocked and dropped calls are cleared. Because of the assumption of exponential inter-arrival times, the system is Markovian; the next system state depends only on the current state.

To specify the model, we define the system events and the calculation of the next event probability. We point out that these next event probabilities are state-dependent. The system events are call arrival, call completion and call crossover. If we assume that the average cellular call duration is t seconds, then the call completion rate for active calls is $\mu=1/t$ calls/second. Further we assume that the average number of crossovers per cellular call is h. Then the cell crossover rate per active call is $h\mu$ crossovers/second. If there are M calls in progress in the whole system, then the system-wide rates for call completion and call crossovers are respectively, $M\mu$ calls/second and $Mh\mu$ crossovers/second.

Next consider the call arrival rates. As a simplification we consider all cells to be statistically identical. (The best fixed assignment scheme assigns an equal number of carriers to each cell.) Since our model is required to achieve large dynamic spatial and temporal variations, we assume that each cell is in one of two modes, DORMANT, characterized by a low call arrival rate λ_{LO} calls/second, and ACTIVE, with a high arrival rate λ_{HI} calls/second. Finally, the model contains two other parameters, N_{HI} and N_{LO}, the average number of call arrivals in the ACTIVE and DORMANT modes, before the cell reverts to the other mode. (The number of call arrivals in each mode is geometrically distributed, with the above means).

The course that the simulation takes depends only on the events and the order in which they occur. As we shall see below (4)-(6), the event probabilities are independent of t. However, the call arrival rates, λ_{HI} and λ_{LO}, and the time spent in the ACTIVE and DORMANT modes, D_{HI} and D_{LO}, depend on t. The time scale of the entire simulation is proportional to t.

In all, the model contains five independent parameters:

h,	average number of crossovers per call;
T_{LO} (Erlangs),	average traffic in the DORMANT mode;
T_{HI} (Erlangs),	average traffic in the ACTIVE mode;
N_{LO},	average number of call arrivals in the DORMANT mode before the cell transitions to the ACTIVE mode;
N_{HI}	average number of call arrivals in the ACTIVE mode before the cell transitions to the DORMANT mode.

The call arrival rates in the two modes are given in terms of the average traffic as, $\quad \lambda_{LO} = T_{LO} \mu$ calls per second $\qquad\qquad$ (1a)

$\qquad \lambda_{HI} = T_{HI} \mu$ calls per second $\qquad\qquad$ (1b)

If the number of cells (instantaneous) in the DORMANT mode is D and in the ACTIVE mode is A, then the system-wide call arrival rate λ is given by,

$$\lambda = D \lambda_{LO} + A \lambda_{HI} \text{ calls per second} \qquad (2)$$

where D, A and λ are the instantaneous (random) values.

The proportion of time that a cell spends in the ACTIVE mode may be considered the ACTIVE mode probability of the cell, p_{HI}. The average duration (in seconds) in the ACTIVE mode is given by $D_{HI} = N_{HI}/\lambda_{HI}$, and the average duration in the DORMANT mode is $D_{LO} = N_{LO}/\lambda_{LO}$. The ACTIVE mode probability is given by,

$$p_{HI} = D_{HI} / (D_{HI} + D_{LO}) \qquad (3)$$

We can now calculate the next-event probabilities. If the instantaneous rates of the events, call arrival (λ), call completion ($M\mu$) and crossover ($Mh\mu$) are known,

Pr{ Next event is Call Arrival} $\quad = \quad \lambda / (\lambda + M\mu + Mh\mu)$ $\qquad$ (4a)

Pr{ Next event is Call Completion} $\; = \quad M\mu / (\lambda + M\mu + Mh\mu)$ $\qquad$ (4b)

Pr{ Next event is Crossover} $\qquad = \quad Mh\mu / (\lambda + M\mu + Mh\mu)$ $\qquad$ (4c)

During the simulation, the next event is determined using the above probabilities. Having determined the next event, we need to determine which cell this event occurs in. Since all calls are considered identical, in the case of a call completion or a crossover, each call has equal likelihood of being selected. Hence if there are M_i calls in progress in cell i, then the probability that the call completion or crossover occurred in cell i is M_i/M. That is

Pr{Call Completion in cell i | Next event is Call Completion} = $\quad M_i/M$ (5a)

Pr{Crossover from cell i | Next event is Crossover } $\qquad = \quad M_i/M$ (5b)

We assume that a crossover from a given cell is to any of the neighboring cells with equal probability (in the highway microcell model, there are only two neighbors).

When the next event is a call arrival, the probability that the arrival is in cell i is given by

Pr{Call Arrival is in cell i | Next event is a Call Arrival} $= \lambda_i/\lambda$ (6)

where $\lambda_i = \lambda_{LO}$ (DORMANT) or λ_{HI} (ACTIVE). At each call arrival, we also determine whether the cell makes a mode transition. The mode transition probability is given by $1/N_i$, with $N_i = N_{LO}$ or N_{HI}, depending on whether the cell is DORMANT or ACTIVE.

4. Dynamic Resource Acquisition Algorithm

To provide distributed control, carrier acquisition decisions are made autonomously at each cell, on the basis of a cost-reward metric calculated from carrier usage information about the neighboring cells. We will elaborate on the size of this neighborhood. The information is collected and updated at each cell based on the knowledge of the system topology (that is the identity of the neighbors) and the carrier acquisition announcements broadcast over the network infrastructure by the base stations. In the cellular packet switch this broadcast information is readily available on the metropolitan area network that links all base stations[1]. In Section 5, we will describe a simplified algorithm, DRA-NN (nearest neighbor), that we have simulated so far. Simulation results for a highway microcell system using the DRA-NN algorithm are given in the next section. Work is currently in progress at WINLAB to further study the DRA algorithm.

A pseudo-code version of the algorithm is in Appendix 1. In the following paragraphs, we first describe the DRA algorithm for a one-dimensional grid of highway microcells, shown in Figure 2. In Figure 1, we show a hexagonal grid of cells (reuse factor = 3), with the DRA algorithm neighborhoods shaded. At the end of this section, we have an example of the cost-reward calculation for hexagonal cells.

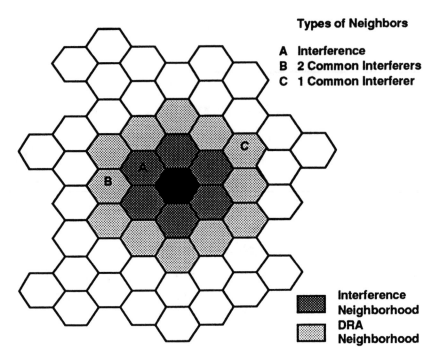

Types of Neighbors

A Interference
B 2 Common Interferers
C 1 Common Interferer

Interference
Neighborhood
DRA
Neighborhood

**Figure 1   Interference and DRA Neighborhoods in Hexagonal
Cells with R=3**

Refer to Figure 2, where the system of microcells is shown. Let us assume that the reuse factor is R, i.e. if a carrier is used in cell i, then the closest cells in which it can be reused are i-R and i+R. (In Figure 2, R=4). At cell i, the carrier usage information for all cells in the range $\{i-2R+2, ..., i+2R-2\}$ must be available. This is defined as the *DRA neighborhood* of cell i. Any carrier that is currently in use anywhere in in the range $\{i-R+1, ..., i+R-1\}$ cannot be used in cell i. We call these cells the *interference neighborhood* of i. For the remaining carriers, a cost metric is calculated and the lowest cost carrier is acquired. The cost associated with *acquiring* a new carrier is the number of cells that are deprived of the use the carrier because of this acquisition. When a carrier is acquired at cell i, it may not be used in the interference neighborhood cells $\{i-R+1, ..., i+R-1\}$. Hence the total number of cells that cannot use the carrier is

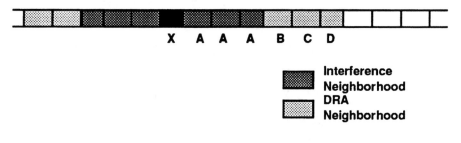

A   Interference
B   3 Common Interferers
C   2 Common Interferers
D   1 Common Interferer

Interference Neighborhood of Cell X
is the set of Cells that may not reuse channels with Cell X
DRA Neighborhood of Cell X
is the set of Cells with Interference Neighborhoods that overlap
with the Interference Neighborhood of Cell X

X   A   A   A   B   C   D

Interference
Neighborhood
DRA
Neighborhood

Figure 2   Interference and DRA Neighborhoods in Highway Microcells with R=4

2R-2. This would be the cost of acquiring the carrier if no other cell in the DRA neighborhood {i-2R+2, ..., i+2R-2} is using the same carrier. Now consider the case when a carrier is in use in another cell in the DRA neighborhood, say cell j. Since some of the cells in the interference neighborhood of cell i are already in the interference neighborhood of cell j, we are not taking away this carrier from those cells. Hence, we must reduce the cost of acquiring the carrier by the number of cells that are already excluded from using the carrier. Thus for any cell j (in the DRA neighborhood of i) where the carrier is in use, we must calculate the number of cells in the intersection of the interference neighborhoods of cells i and j. We must subtract this number from the total cost of using the carrier in cell i. For example, if the carrier is in use at cell i+R+2, then its interference neighborhood is {i+3, ..., i+2R+1}. Hence, the set {i+3, ... i+R-1} is the intersection of the two interference neighborhoods and its cardinality is R-3. The cost of acquiring the carrier must be reduced by this amount from 2R-2. This gives the acquisition cost for this carrier as 2R-2 - (R-3) = R+1.

Also notice that if a carrier is in use at the two co-channel neighbors i-R and i+R, then the intersections are such that the cost of using the carrier at cell i is zero. This is as it should be to ensure maximum packing.

Let us next consider carrier release by a cell. Clearly this should be done so as to release the carrier that frees up its use in the maximum number of cells. Instead of finding the minimum cost we now consider maximizing the reward from the carrier release, where the reward is the number of cells at which the carrier will be available for use if it is released by cell i. For example, notice that if the carrier is in use at i-R and i+R, then releasing the carrier at cell i, does not make the carrier available at any of the cells in the interference neighborhood of i. Thus the reward from releasing this carrier would be zero. The reward is greatest (2R-2) if the carrier is not in use in any of the cells in the DRA neighborhood of cell i. If the carrier is in use in cell j in the DRA neighborhood of cell i, then the reward must be reduced by the number of cells in the intersection of the interference neighborhoods of cells i and j.

Notice that the calculation of the cost and the reward in the two cases above is the same. The difference is that in case of carrier acquisition, we want to take the carrier away from the fewest number of cells and so we are looking for the carrier that has the *lowest* cost. Thus, we are looking for a carrier that is in use at cells whose interference neighborhoods have the largest intersection with the interference neighborhood of the current cell. On the other hand, for carrier release, we are looking for a carrier that is in use at cells whose interference neighborhoods have the smallest intersection with interference neighborhood of the current cell. This opens up the carrier for use in the greatest number of cells in the interference neighborhood of the current cell and therefore has the *highest* reward.

Example: Consider the system of hexagonal cells with reuse=3 shown in Figure 1. In Figure 3, the cell under consideration is shaded black. Let us call it

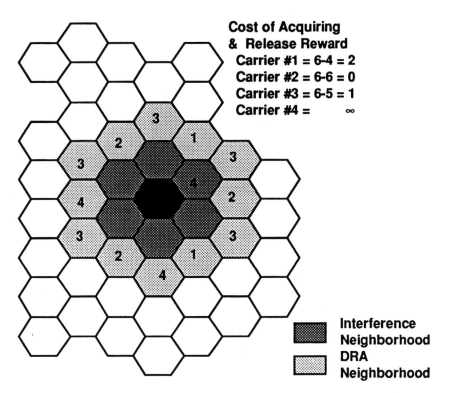

Figure 3   Cost and Reward calculation for Hexagonal Cells

cell X. The state of carrier acquisitions at all the cells in the DRA neighborhood of cell X is as shown. By considering each of the Carriers #1, #2, #3 and #4, we can determine the cost and reward associated with acquiring/releasing the carrier at cell X.   With a reuse of 3, the interference neighborhood consists of 6 cells. Hence INsize = 6. Four of the six cells in the interference neighborhood of cell X are already deprived of using carrier #1. Hence, the cost/reward for carrier #1 at cell X is 2.   Similarly, the cost/reward for carriers #2 and #3, are 0 and 1, respectively. Hence, the carriers are rank-ordered, #2, #3, #1, #4. If a carrier is to be acquired at cell X, it acquires the first carrier in the above rank-ordered list, that is not already in use at cell X. To release a carrier, cell X must release the last carrier in the above list, that is in use at cell X.

Since carrier acquisition and release cannot be performed instantaneously, to keep call blocking and dropping small we must always maintain a buffer of B voice circuits (TDMA slots) at each cell (if possible). When fewer than B slots remain unused, the carrier acquisition procedure begins. Similarly, if the TDMA carriers consist of N slots (voice channels), then a carrier release procedure begins when more than N+B slots are unused. This system parameter B must be determined based on the time constants of the DRA algorithm as well as the dynamics of the traffic. The propagation time of the carrier acquisition and release information over the network infrastructure is also important in determining B, but for the present we have assumed instantaneous information availability over the wireless network infrastructure.

5. Results of Highway Microcell Simulation

We have simulated the simplified resource acquisition algorithm, DRA-NN, operating in a highway microcell system. For DRA-NN, the cost/reward calculation at each cell takes into account the closest reuse neighbors. In any cell, it is preferable to use carriers that are also in use at the cell's reuse neighbors. Therefore the carriers used in the reuse-neighbor cells are the first to be acquired (lowest cost), and the last to be released (lowest reward). The only information required at each cell is the carrier usage in the interference neighborhood. This is a substantial reduction from the DRA algorithm since the interference neighborhood is a small subset of the DRA neighborhood for a cell, especially in two dimensions (Figure 1). Early work taking an approach similar to DRA-NN [11,12] suggests that this simplified algorithm will also provide good performance.

Our aim is to explore the ability of the technique to rearrange network resources in response to changing teletraffic patterns. Relevant performance measures are the probability of call blocking for arriving calls and the probability of call

Number of cells	25
Number of TDMA carriers	20
Slots per TDMA carrier (N)	20
Reuse factor (R)	5

FIXED ALLOCATION

4 TDMA carriers per cell

Capacity: up to 80 simultaneous conversations per cell

DRA

20 TDMA carriers

Capacity: up to 400 simultaneous conversations in 5 contiguous cells

Buffer	BUFF	5 voice channels

MICROCELL CHARACTERISTICS

Traffic (in ACTIVE cells)	T_{HI}	200 Erlangs
Traffic (in DORMANT cells)	T_{LO}	20 Erlangs
Average Traffic	T_{AV}	40 Erlangs
Average number of call arrivals		
(in ACTIVE mode)	N_{HI}	2000 calls
(in DORMANT mode)	N_{LO}	1600 calls
Probability cell is ACTIVE	p_{HI}	1/9
Duration in ACTIVE mode	D_{HI}	15 minutes

CALL CHARACTERISTICS

Average call duration	t	90 seconds
Average number of crossovers per call	h	2.0

Table 1. Nominal Values of System Parameters for the Highway Microcell Simulation of DRA-NN

dropping at handoff. For the simulation we calculate the system-wide blocking probability as the ratio of the number of blocked calls to the total number of call arrivals. The call dropping probability is calculated as the ratio of the number of calls dropped at a cell boundary crossover, to the total number of cell boundary crossovers. Starting with the nominal traffic conditions listed in Table 1, we varied several activity parameters and in each case measured call blocking and call dropping probabilities for DRA-NN and fixed channel allocation (FCA). In each case DRA-NN outperformed FCA.

Figure 4 shows the effect of varying the geographical uniformity of traffic. (Notice that we have used slightly different values of parameters for Figure 4, than those listed in Table 1). On the horizontal axis is the ratio of the call arrival rate in an ACTIVE cell to the call arrival rate in a DORMANT cell. When this

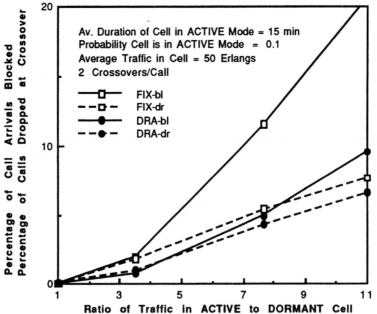

Figure 4

ratio is unity, the traffic is uniform. The blocking and dropping probabilities are very low because the average offered traffic 50 Erlangs per cell, is handled with low blocking by FCA with 80 voice channels per cell. With non-uniform traffic, some cells experience demand far exceeding the 80 available channels, and both the blocking probability (solid curves) and the dropping probability (dashed curves) increase. However, the rate of increase with FCA (square data points) is higher than the rate of increase with DRA-NN (circle data points).

In Figure 5-7, we use the parameters from Table 1, and vary one parameter at a time. In Figure 5, the ACTIVE/DORMANT traffic ratio is fixed at 10:1 and the amount of time that a cell spends in the ACTIVE mode is varied. The disparity between the ACTIVE and DORMANT cells increases as this duration increases. Figure 5 shows an increasing advantage of DRA-NN over FCA, with increasing

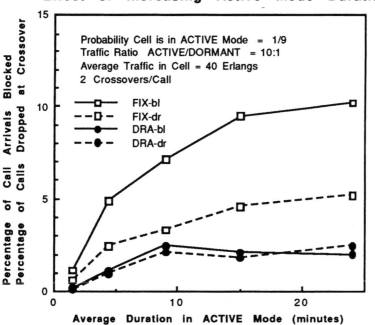

Figure 5

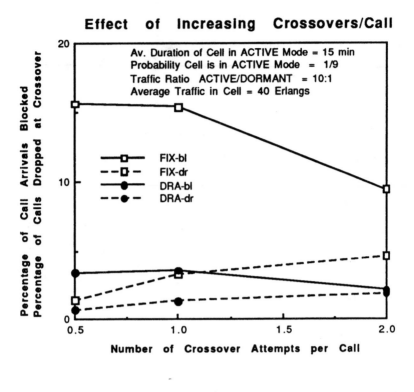

Figure 6

amount of time spent by cells in the ACTIVE mode.

In Figure 6, we explore the effect of changing user mobility, as measured by the number of cell boundary crossings per call. As this number increases more handovers are required and the call dropping probability (calculated from the number of unsuccessful handover attempts) increases. However, the call blocking probability actually decreases because mobility tends to take calls from ACTIVE cells to neighboring DORMANT cells, thus freeing up channels in the ACTIVE cells for new calls. Thus the increased mobility has the effect of making the overall traffic less non-uniform.

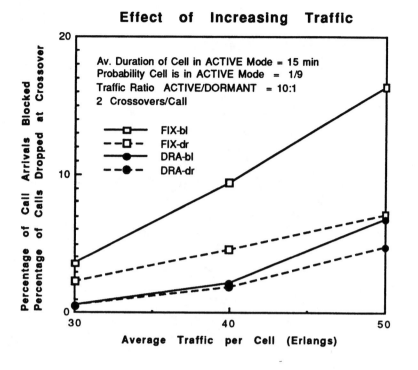

Figure 7

Finally, in Figure 7, we consider the effect of increasing the total system-wide traffic, while keeping the ACTIVE/DORMANT traffic ratio at 10:1. In the plot we have shown the results as the average offered load per cell is increased from 30 to 50 Erlangs. We see that in this range DRA-NN outperforms FCA. As we increase the traffic beyond this, we may encounter the point beyond which FCA has lower blocking than DRA-NN. However, since at the average offered load of 50 Erlangs per cell, we already have FCA blocking greater than 15%, the range of higher offered traffic is uninteresting from the system design point of view.

To conclude, we repeat some of the open issues mentioned earlier. Simulations of the DRA algorithm in two dimensions must be performed and compared with the DRA-NN performance. We have ignored the time delays involved in each step of the algorithm. This includes the delay in the channel usage information being provided to the neighbors and the delay in calculation of the cost/reward metric and subsequent channel acquisition/release. Since the algorithm responds to the traffic stimulus that is distributed over the different cells, the algorithm performance would deteriorate substantially if these delays are long relative to the rate of change of the traffic patterns. In addition to further study on several aspects of the DRA algorithm, more work is needed to validate and refine the traffic model, and to identify observable parameters relevant to resource acquisition.

Acknowledgment

We would like to acknowledge useful discussions on the topic with the Chris Rose, John MacLellan and Kathy Meier-Hellstern.

References

1. David J. Goodman, "Cellular Packet Communications," *IEEE Transactions on Communications*, vol. 38, no.8, pp. 1272-1280, August 1990.

2. Raymond Steele, "The Cellular Environment of Lightweight Handheld Portables," *IEEE Communications Magazine*, pp. 20-29, July 1990.

3. D.J. Goodman, R.A. Valenzuela, K.T. Gayliard, and B. Ramamurthi, "Packet Reservation Multiple Access for Local Wireless Communications," *IEEE Transaction on Communications*, vol. 37, no. 8, pp. 885-890, August 1989.

4. Sanjiv Nanda, "Analysis of Packet Reservation Multiple Access: Voice-Data Integration for Wireless Networks," *Proc. Globecom '90*, pp. 1984-1988, San Diego, December 1990.

5. A. Salmasi, "An Overview of CDMA applied to the Design of PCN's," *Proceedings of the Second WIN Workshop*, in this volume.

6. R. Beck and H. Panzer, "Strategies for Handover and Dynamic Channel

Allocation in Micro-Cellular Mobile Radio Systems," *Proc. IEEE Vehicular Tech. Conf.*, May 1989.

7. Y. Furuya and Y. Akaiwa, "Channel Segregation, A Distributed Adaptive Channel Allocation Scheme for Mobile Communication Systems," *Proc. Second Nordic Seminar on Digital Land Mobile Communications*, Oct 1986.

8. T. J. Kahwa and N. D. Georganas, "A Hybrid Channel Assignment Scheme in Large-Scale Cellular-Structured Mobile Communication Systems," *IEEE Transactions on Communications*, vol. 26, no. 4, pp. 432-438, April 1978.

9. D. Everitt and D. Manfield, "Performance Analysis of Cellular Mobile Communication Systems with Dynamic Channel Assignment," *IEEE Journal on Selected Areas in Communication*, vol. 7, no. 8, pp. 1172-1180, Oct. 1989.

10. K. N. Sivarajan, R. J. McEliece and J. W. Ketchum, "Dynamic Channel Assignment in Cellular Radio," *Proc. IEEE Vehicular Tech. Conf.*, pp. 631-635, May 1990.

11. D. C. Cox and D. O. Reudink, "Effects of some Nonuniform Spatial Demand Profiles on Mobile Radio System Performance," *IEEE Transactions on Vehicular Tech.*, vol. 21, no. 2, pp. 62-67, May 1972.

12. D. C. Cox and D. O. Reudink, "Dynamic Channel Assignment in Two-Dimensional Large-Scale Mobile Radio Systems," *Bell System Technical Journal*, vol. 51, no. 7, pp. 1611-1629, September 1972.

Appendix 1 The DRA Algorithm

I INITIALIZE

Reinitialize on changes in the system topology and other long term changes in average traffic.

For each cell i in the system, the following quantities are stored at each cell site at Initialization. This information will be used locally by the DRA algorithm to make carrier *acquisition* and carrier *release* decisions.

Definitions:

Interference neighborhood IN_i = set of all cells j, such that if a carrier is in use at cell j, then it cannot be reused in the cell i

DRA neighborhood $DRAN_i$ = cell $j \in DRAN_i$, if $IN_i \cap IN_j \neq \phi$ that is, if the interference neighborhoods of cells i and j intersect, then cell j is in the DRA neighborhood of cell i.

Maximum cost $INsize_i$ = $Card(IN_i)$ This is the maximum number of cells deprived of a carrier acquired at cell i

TDMA slots per frame N = Number of voice channels per TDMA carrier

Number of carriers $CARR$ = Maximum number of TDMA carriers available for DRA

Buffer Size (slots) $BUFF$ = Minimum number of slots to be kept available at any cell (if permitted by current TDMA carrier usage at other cells). This determines when carrier *acquisition* and *release* operations are begun.

At Initialization:

Download to cell site i, the above quantities:
    system variables, $N, BUFF, CARR$
    cell sets, $IN_i, DRAN_i$

II CARRIER USAGE UPDATES

Carrier Usage Updates are performed at each acquisition/release announcement received over the fixed network *DRA-Broadcasts* of the form:

$$\text{acquisition/release; carrier } K; \text{ cell } J$$

Definitions:

Carrier Usage matrix $USAGE_i(j,k)$ $=$ 1   if carrier k is in use at cell j

    0   if carrier k is not in use at cell j
$$j \in DRAN_i \, , \, 1 \le k \le CARR$$
carrier usage information for cell i

Carrier Availability $AVAIL_i(k)$ $=$ 0   if carrier k is in use in IN_i

    1   otherwise
Availability of carrier k at cell i

Common Interference set $CI_i(k)$ $=$ $j \in CI_i(k)$, if $AVAIL_i(k) = 1$
$$\text{and } USAGE_i(j,k) = 1$$
The set of cells in $DRAN_i - IN_i$
at which carrier k is in use.

Number of Deprived cells $DEP_i(k)$ $=$ -Infinity   if $AVAIL_i(k) = 0$

Card($\cup_{j \in CIi(k)} \{ IN_j \cap IN_i \}$)
$(= 0$, if $CI_i(k)$ is empty$)$

Number of cells in IN_i already

deprived from using carrier k

Carrier Release Reward $REWARD_i(k) =$ $INsize_i - DEP_i(k)$

Carrier Acquisition Cost $COST_i(k)$ $=$ $INsize_i - DEP_i(k)$

At each DRA Broadcast Message Received at Cell i:

    IF $J \in DRAN_i$ THEN   Update $USAGE_i(J,K)$

                                  $AVAIL_i(K)$

                                  $CI_i(K)$

               Calculate $DEP_i(K)$

                                  $REWARD_i(K)$

                                  $COST_i(K)$

    ENDIF

III CARRIER ACQUISITION & RELEASE

Carrier acquisition/release actions are performed at each cell i, autonomously, using the information available locally, as defined above. The process is subject to the constraint that at least *BUFF* slots must be available at the cell i, if the carrier usage in $DRAN_i$ permits.

Definition:
Slots Unused $UNUSED_i$ = Remaining number of slots available at cell i

At each Call Arrival, Completion or Crossover at Cell i:
IF $UNUSED_i \leq BUFF$ THEN

        OVER ALL k with $USAGE_i(i,k) = 0$,

            Find K such that $COST_i(K) \leq COST_i(k)$ for all $k{\neq}K$

        ENDOVER
        IF $COST_i(K) \leq INsize_i$ THEN

            Acquire carrier K
            Transmit DRA Broadcast: acquisition; carrier K; cell i

        ENDIF
ENDIF

IF $UNUSED_i \geq N + BUFF$ THEN

        OVER ALL k with $USAGE_i(i,k) = 1$,

            Find K such that $REWARD_i(K) \geq REWARD_i(k)$ for all $k{\neq}K$

        ENDOVER
        IF $REWARD_i(K) \geq 0$ THEN

            Release carrier K
            Transmit DRA Broadcast: release; carrier K; cell i

        ENDIF
ENDIF

On Dynamic Channel Allocation In Cellular/Wireless Networks

Srikanta P. R. Kumar and **Hwan Woo Chung**
Electrical Engineering and Computer Science
Northwestern University
Evanston, Illinois 60208

Mohan Lakshminarayan
Bell Northern Research
Richardson, Texas 75081

Abstract

With the trend towards higher network performance in cellular and wireless networks, traffic sensitive methods that assign available channels to cells (macro and micro) more dynamically (instead of fixed assignment) may prove beneficial. In general, dynamic channel allocation methods can generate better capacity and better handoff performance (lesser forced terminations). Previous studies of dynamic methods are largely simulation based, and theoretical studies aimed at a better understanding of the benefits are relatively few. In this paper, we study several dynamic channel allocation methods, including channel borrowing and hybrid techniques, and analyze them in a stochastic framework. Comparison of allocation methods using stochastic dominance concept is illustrated. This facilitates a finer comparison than the conventional approaches. Bounds indicating the conditions under which dynamic methods considered dominate fixed allocation methods are also developed. Simulation studies are also reported.

1 Introduction

Typical design philosophy in cellular and wireless communication networks is to assign the available channels in a fixed manner to each cell, with spatial reuse. Co-channel interference considerations and the cellular geometry dictate the size and structure of reuse clusters. The conventional approach to increasing capacity is via cell splitting which facilitates a higher degree of spatial frequency reuse. With the anticipated rapid growth in cellular and wireless network traffic, and the trend towards higher network performance, traffic sensitive methods that assign available channels to cells more dynamically, instead of fixed assignment, may prove beneficial. In general, dynamic channel allocation methods exhibit better call carrying capacity, and also better handoff performance in terms of lesser forced terminations. With the current trend towards digital systems with their associated flexibility, and increased intelligence and capabilities in terminals and base stations, sophisticated dynamic channel allocation methods are an attractive way to improve network capacity and performance. Furthermore, as some recent studies indicate, dynamic allocation may be necessary in many instances. For example, in micro-cellular system (high density, interference limited, irregular zones), spatial volatility of traffic is likely to be high, and therefore dynamic methods may be virtually essential. In this paper, we investigate some properties of dynamic channel allocation methods.

There is a considerable literature on dynamic channel allocation. However, most of these studies use simulation tools for analysis, and in addition, handoff considerations are usually ignored. Theoretical studies aimed at a better understanding of the benefits of dynamic methods are relatively few. While simulation is versatile, it is also often less insightful. Theoretical studies based on mathematical models, though often require restrictive assumptions, can provide useful insights. In this paper, we study several dynamic channel allocation methods, including channel borrowing and hybrid techniques, and analyze them in a stochastic framework. Comparison of allocation methods using stochastic dominance concept is illustrated. This facilitates a finer comparison than the conventional approaches. Bounds indicating the conditions under which dynamic methods considered dominate fixed allocation methods are also developed. Simulation studies are also reported.

The paper is organized as follows. In section 1.1, we briefly discuss some general aspects of dynamic allocation, and in section 1.2, previous work is briefly reviewed. Section 2 describes the model considered. In section 2.1, we illustrate the use of stochastic dominance in comparing the call carrying capacity of channel borrowing schemes and fixed allocation, ignoring co-channel interference consideration. Section 2.2 extends the analysis to incorporate spatial reuse and interference effects, and to derive sufficient conditions for dynamic methods considered to perform better than fixed allocation. Some simulation results are presented in section 3.

1.1 Dynamic Channel Allocation Strategies

In contrast to fixed assignment methods, dynamic channel allocation allows for assignment on demand, thereby responding to call/requests more dynamically. This is provided typically via some form of channel sharing or pooling among cells. The extent of pooling, however, may differ from one method to another. With multiple user or service classes (e.g., integrated services, analog/digital subscribers, etc.), channels are dynamically shared across classes. Various dynamic allocation methods in cellular systems may be derived based on a number of considerations. Some of these factors, and a rough classification are given below.

Spatial Boundaries for Channel Sharing : In the simplest case, channel sharing is limited to adjacent cells. In other words, each cell is allowed to borrow or lend a channel only to its neighboring cells. The number of shared channels may also be limited to some fixed quantity. More generally, channel sharing or pooling boundaries may be extended to k-nearest neighbors ($k > 1$), to some specified clusters, or even to the entire area of cellular coverage.

Hybrid Allocation (some fixed, some shared) : Some of the available channels are fixed for each cell (with reuse clusters for these determined from usual considerations), whereas a certain number of channels are available for sharing.

Allocation Dependent on State/Traffic : While the above specify the region for sharing and the number of channel shared, they do not specify when a channel should be shared. This decision could be based on a number of factors. For example, the states or the number of calls in progress (or even some function of carried or offered traffic) in adjacent cells is one factor. More generally, the dynamic allocation could be based on the states of adjacent clusters or the states of some well defined relevant region (in which the impact of the decision will be most felt).

Other considerations may include allocation across user/service classes [9], or allocation in a hierarchical architecture. With macro cells overlaid on micro cells, channel sharing could be restricted to within the same level of hierarchy, or sharing could be permitted across the levels.

1.2 Previous Work

Some of the early work on dynamic channel allocation in cellular systems was done by Cox and Reudink [1]. They considered certain types of dynamic assignment (predominantly, one method where the shared boundary is the entire area of coverage), and in a simulation study showed the qualitative benefits of such methods in 1-D (highway, major air-route) and 2-D geometries. Schiff [2] studied one method of dynamic allocation in an analytical framework; however, this was again for simple systems with one user class, and handoff considerations are ignored. Choudhary and Rappaport [10], and Hong and Rappaport [11] study dynamic allocation, primarily for improving handoff performance for single class. Xu and Mirchandani [4] study adjacent channel sharing (borrowing and lending) focusing on a single class not considering handoffs. See also Everitt and Manfield [3] for a recent study in the context of small-cell systems. Others who have addressed this issue include [12]-[17].

There is also a considerable body of literature developed in the context of traditional telephone networks and associated capacity issues (see [5]-[9] for some recent analytical studies). These are not directly applicable in the present context, as the cellular system features are different.

2 Network Models and Channel Allocation

In this section, we formulate and analyze simple mathematical models of a cellular network and dynamic channel allocation methods. We examine several systems to get more theoretical insight, and also to characterize the conditions and estimates of the parameters, for which dynamic methods are superior to fixed channel allocation. The motivation is that such a study will be a guide for more complex systems, and also for simulation studies. We focus on the comparison of different systems arising from different allocation methods. Particular attention is paid to call carrying capacity; other performance measures such as blocking probability of first-attempt calls and probability of forced termination will be discussed. The notion of stochastic dominance, which is useful in this context is reviewed next. In section 2.1, we illustrate the comparison technique by examining a channel borrowing scheme in a simple example ignoring co-channel interference consideration. In section 2.2, extension to include interference effects is considered.

Definition 1 *A random variable X stochastically dominates another random variable Y, denoted $X \geq_s Y$ if $Pr[X > \alpha] \geq Pr[Y > \alpha]$, for all α.*

In the analysis of the network models considered later, the random variables will be the calls carried (or the state at a certain instant or in steady state) in two different systems. The above definition implies that if F_X and F_Y are respectively the distribution of calls carried in two systems, then $F_X(\alpha) \leq F_Y(\alpha)$ for all α, if $X \geq_s Y$.

One of the advantages of using the above notion of comparison (see Ross [18]) is that for any revenue function $r(\cdot)$ which is increasing, the above definition is equivalent to the condition

$$E[r(X)] \geq E[r(Y)]. \tag{1}$$

By choosing different revenue functions, different measures of performance can be considered. If the random variable denotes the number of calls carried, then taking r as identity yields the expected call carrying capacity comparison via (1).

In the Markov models of networks considered, the following technique will be used to generate stochastic dominance results. Given two systems $i = 1, 2$, let X_t^i be the state of system i at time t. To show system 1 is better than system 2, we shall assume dominance at time t (i.e., $X_t^1 \geq_s X_t^2$), and show that the network evolution is such that the same dominance holds at time $t + dt$ also. This would, then, imply that given that both systems are started up with the same initial conditions, one system will be stochastically better than the other at each point in time, and also the expected revenues generated will be higher. This result is a stronger form of comparison than the usual measures of comparison via steady state values.

2.1   A Simple Illustration

To illustrate the analytical technique, a simple example is considered here ignoring co-channel considerations. Consider the seven cell system shown in Fig. 1. Suppose there are N distinct channels assigned per cell. Traffic (calls) originates in each cell according to a Poisson process with rate λ. We abstract away from cell design considerations. Assume that each cell is of equal size, and that traffic is spatially homogeneous. Assume there is a single service class, and each call connected (or in progress) hangs up at constant hazard rate μ; and the handoff to an adjacent cell also has a constant hazard rate δ. This implies both conversation and handoff times are exponential, and a call successfully handed off may be viewed as a new call (due to the

130

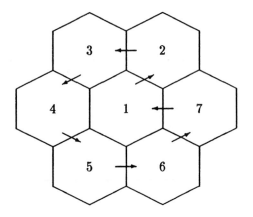

Figure 1: *7-cell structure*

memoryless property). Assume that the handoff directions are uniform in space (six directions), and the cells are wrapped around. Call set-up and handoff processing times are assumed negligible. These assumptions, though somewhat simplistic, have been used in the literature, and are a good starting point for analysis.

In the above example, if the handoff rate δ equals zero, then the steady state blocking probability and expected calls carried are readily computed from the Erlang's formula. With strictly positive handoff rates, the analysis could get quite complex, particularly in large networks. In the following, however, we take a different tack in comparing fixed channel allocation with a simple class of channel borrowing schemes.

Suppose that a cell can borrow (on demand when all of its current channels are busy) up to k channels from m other cells, in a symmetric pattern. An example pattern for $m = 1$ is shown via the arrows in Fig. 1; cell 1 can borrow from cell 7, cell 2 can borrow from cell 1 and so on. It should be pointed out that the cells need not be hexagonal; they could be rectangular or any other shape.

Let X_t^i denote the number of calls carried at time t for $i = b$(borrowing scheme) or f(fixed allocation). The general idea of the comparison is to consider the time interval $[t, t + dt]$, and first show that, in this interval, the probability of an upward transition is higher and that of a downward transition is lower for the borrowing scheme than for fixed allocation. This property can then be used for characterization via (1). The downward and upward transition probabilities are shown in Fig. 2, for $X_t^i = x$. ($o(dt)$ term has been

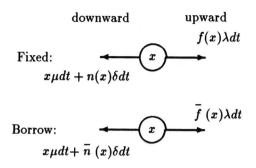

Figure 2: *Transition Probabilities in the Example*

omitted, as it is harmless.) The borrowing schemes considered generate state transitions such that

$$\bar{n}(x) \leq n(x), \qquad \text{and,}$$

$$\bar{f}(x) \geq f(x) \qquad \text{for all } x, \tag{2}$$

where $n(x)\delta$ and $\bar{n}(x)\delta$ are rates of forced terminations due to handoffs. To illustrate (2) for $N = 1$, that is, one channel per cell, consider the borrowing pattern shown in Fig. 1. For fixed assignment, $n(x) = x(x-1)$ for $x = 0, \cdots, 7$. However, for borrowing $\bar{n}(x)$ is equal to $n(x)$ or takes a value less than $n(x)$ which depends on the pattern of the calls in progress, and the borrowing pattern. For example, in the system with $N = 1$, let $x = 3$. Suppose there are calls carried in cells 1, 2 and 7; then $\bar{n}(x) = 2$, since handoffs from cell 2 to cell 1, and from cell 7 to cell 2 will result in forced termination . But for the case when calls are residing in cells 1, 4 and 7, $\bar{n}(x) = 1$, since the only forced termination from this state is a handoff from cell 4 to cell 1. And finally, for the calls residing in cells 1, 2 and 6, $\bar{n}(x) = 1$. In all cases, (2) is satisfied.

We wish to point out that even for the non-wrap around structure, (2) holds. With assumptions as above, the following hold. For calls in 1, 2 and 7, $\bar{n}(3) = 2$; for calls in 1, 4 and 7, and for calls in 1, 2 and 6, $\bar{n}(3) = 0$.

Given the relationship (2), it can be established that in this model, borrowing schemes are better than fixed allocation, in the sense explained in previous section. Define for $i = b, f$,

$$Pr[X_{t+dt}^i \geq x \mid X_t^i = y] \equiv t^i(y; x)$$

The following table lists the above transition probabilities.

y:	0	1	2	$\cdots$	$x-1$	x	$x+1$	$\cdots$
$t^f(y;x)$:	0	0	0	$\cdots$	$f(x)\lambda dt$	$[1-x\mu dt - n(x)\delta dt]$	1	$1\cdots$
$t^b(y;x)$:	0	0	0	$\cdots$	$\bar{f}(x)\lambda dt$	$[1-x\mu dt - \bar{n}(x)\delta dt]$	1	$1\cdots$

Two properties in the above table are important.

$$(i) \quad t^b(y;x) \geq t^f(y;x), \text{ for all } y \text{ and } x \qquad (3)$$

$$(ii) \quad t^i(y;x) \text{ is increasing in } y, \text{ for } i = f, b. \qquad (4)$$

Now for every state x,

$$\begin{aligned}
Pr[X^b_{t+dt} \geq x] &= \sum_y t^b(y;x) Pr[X^b_t = y] \\
&= E[t^b(X^b_t;x)] \\
&\geq E[t^b(X^f_t;x)] \\
&\geq E[t^f(X^f_t;x)] = Pr[X^f_{t+dt} \geq x] \qquad (5)
\end{aligned}$$

The first inequality follows from the application of (1), property (4), and the assumption $X^b_t \geq_s X^f_t$. The second inequality follows from (3). Relation (5) establishes :

Proposition 1 *In the model (of no co-channel interference) above, if the systems with borrowing and fixed channel allocation schemes start in the same state, then for all $t \geq 0$, $X^b_t \geq_s X^f_t$.*

In other words, calls carried is stochastically larger at all times with borrowing scheme than with fixed allocation. The same result holds in steady state (initial state does not matter as both systems are ergodic). Furthermore, for any increasing revenue function, expected revenues are better with the borrowing scheme. It can also be shown that the rate of successful handoffs and connection of first attempt calls will be better. For example, to show handoff completions are better, define the rates $r_b(x) \equiv 6x\delta - \bar{n}(x)$ and $r_f(x) \equiv 6x\delta - n(x)$. It is easily verified that, for $x = 0, 1, \cdots, 7$, $r_f(x)$ equals $0, 6\delta, 10\delta, 12\delta, 12\delta, 10\delta, 6\delta$, and 0; and $r_b(x)$ is greater than or equal to $0, 6\delta, 12\delta, 16\delta, 19\delta, 15\delta, 11\delta$, and 7δ. It is, then, clear that there exists an increasing function $r_0(x)$ such that

$$r_b(x) \geq r_0(x) \geq r_f(x). \qquad (6)$$

Let C_i be the expected success rate of handoffs for $i = b, f$. Then

$$
\begin{aligned}
C_b &= \sum_x r_b(n) Pr[X^b = x] \\
&= E[r_b(X^b)] \\
&\geq E[r_0(X^b)] \\
&\geq E[r_0(X^f)] \geq E[r_f(X^f)] = C_f.
\end{aligned}
\tag{7}
$$

The first and the third inequality (7) follows from (6), and the second inequality is due to $X^b \geq_s X^f$ and the increasing property of $r_0(x)$.

With regard to the call completion, let $\bar{C}(x) \equiv x\mu$ denote the rate of completion (calls leaving the network after they are served) when there are x calls in progress. Since $\bar{C}(x)$ is increasing in x, it follows that, at all times, the expected call completion rate with the borrowing scheme $\bar{C}_b \equiv E[\bar{C}(X_t^b)]$ is higher than that for fixed allocation, $\bar{C}_b = E[\bar{C}(X_t^f)]$. Since blocking rate of first attempt calls equals the total arrival rate minus the completion rate, it follows that the borrowing scheme has lesser first attempt blocking.

Proposition 2 *In the above model, the success rates of handoffs and first attempts are higher for the borrowing scheme at all times.*

Moreover, in this example, the following result also holds (proof omitted here).

Proposition 3 *In the model above, calls carried increases stochastically in both k and m.*

Similar results can be derived for other borrowing schemes, for the above model. In fact, these results are related to the well known concept of trunking efficiency. They characterize this efficiency in stronger terms of stochastic dominance, and also incorporating handoff considerations. The next section extends this analysis to include co-channel considerations.

2.2 Model with Spatial Reuse

Co-channel interference is a primary consideration in spatial reuse of channels. In this section, we illustrate how co-channel effects can be incorporated into the analysis outlined in section 2.1. Here, we focus on a particular class of dynamic allocation viz.,, hybrid borrowing channel assignment method, and show how one can ascertain conditions for which channel borrowing is better than fixed allocation. The general technique of comparison is applicable for analyzing other dynamic allocation methods as well.

We retain the traffic (call origination, completion, and handoff) assumptions of section 2.1, with N channels per cell. Suppose the reuse cluster size is seven (2-cell buffering); hence the total number of available channels is $7N$.

The hybrid allocation considered is a mixture of fixed allocation and channel borrowing on demand [12][13]. Suppose the N channels of each cell are divided into two sets : a fixed set of channels dedicated to the cell, and a set its adjacent cells can borrow. Let A_i denote fixed set of $N - k$ channels, and B_i denote set of k borrowble channels. Since the reuse cluster size is seven, we need to identify only $B_1, \cdots, B_7$. The method corresponds to a fixed frequency plan, i.e., cell i is designed to handle the frequencies $A_i \cup [\bigcup_{i=1}^{7} B_i]$.

The borrowing strategy is as follows (see [14] and [15]). A cell, on demand, (to handle a call when all channels currently in it are busy) borrows a channel from the B-set of a neighbor. Highest priority is given to a neighbor with maximum B-set availability. Once a channel is borrowed, co-channel cells in the adjacent clusters are locked, i.e., the channel cannot be used in those cells (with seven cell reuse pattern, there are two adjacent clusters that are affected).

A cell i releases a borrowed channel (to its original or home cell), if a channel from its fixed set A_i or its own B_i is available, by switching the call. If there is a choice of which borrowed channel to release, priority may be assigned on the basis of first-in-first-out or the neighbor (home cell) which is most busy. Several variations around this borrowing scheme are possible and have been reported [14][15][16]. The precise details of the borrowing scheme are not necessary in the analysis (to follow) of the condition for which borrowing dominates fixed allocation.

In practice, the above model of a cellular system can be analyzed by conventional Markov chain techniques. However, seeking solution (to find expected call capacity etc.) via this straight forward approach of finding the steady state probabilities of the Markov chain (numerically or via simulation) could become an unwieldy task, especially in large networks (large number of clusters, channels etc.), due to very large number of states needed to characterize the system.

In the spirit of seeking qualitative insights, we examine, in the following, conditions under which borrowing schemes do better than fixed allocation. Our starting point is a single reuse cluster considered in section 2.1; to this, we attempt to incorporate factors reflecting the co-channel effects of borrowing in a way that enables the comparison of the allocation methods via stochastic dominance. As we wish to keep the extension simple, the analysis presented is approximate (refinements are possible), and the bounds derived are conservative.

The approach taken here is to examine the positive and negative effects of borrowing a channel. Consider the moment at which cell i wants to borrow a channel from cell j. One positive aspect to borrowing the channel is the completion of this call at hand in cell i. There are two negative factors : first, a potential (future) call in cell j may be lost due to channel lending; and second, potential calls in the co-channel cells (say, N_c in number) in the clusters adjacent to that of cell i, may be lost due to locking. The main implication of the analysis of section 2.1 is that, in the absence of co-channel interference, the net effect to borrowing (positive minus the first negative effect) is positive. Taking the somewhat simplified picture of all the effects as described above, the expected benefits from a single call on a borrowed channel, denoted V may be determined as follows.

Suppose cell i wants to borrow from cell j at time t. Let k_0 represents the number of busy channels in cell i at this instant. Considering a short time interval $[t, t+dt]$, the probability a call completion occurs (approximately, the probability of returning this borrowed channel) is $(k_0 + 1)\mu dt$. This may be viewed as generating a unit reward. A handoff from cell i to cell j in this interval, which has probability $(k_0 + 1)\delta dt$, may be seen as resulting in the return of the borrowed channel. A handoff to any other cell $k(\neq j)$, which occurs with probability $(k_0 + 1)5\delta dt$, may or may not result in borrowing by cell k. It is conservative to assume that it does, and we do so. In this case, the expected benefits continues to be V. In the event, the status of the borrowed channel in cell i does not change (see third term on the right side of (8) below), no gains are accrued during dt, and the expected future benefits remain at V. On the negative side, the expected penalty paid during this interval is $N_c \lambda dt P$, where P denotes the probability of attempt in one of the N_c co-channels in the adjacent clusters. Hence

$$V = (k_0+1)(\mu+\delta)dt + (k_0+1)5\delta dt V + [1-(k_0+1)(\mu+6\delta)dt]V - N_c\lambda dt P + o(dt) \tag{8}$$

Manipulation of (8) yields

$$V = 1 - N_c\lambda P/(k_0 + 1)(\mu + \delta). \tag{9}$$

If $V > 0$, borrowing a channel generates a positive revenue flow. The entire cellular system may now be viewed clockwise, by taking appropriate values for the parameters in (9). In particular, P and k_0 need to be determined. Assessing their values precisely in a system with large number of cells (from the Markov model of the entire network) is involved. A simple approximation is as follows. A conservative value of P is $E_r(\lambda/\mu; N-k)$, where E_r is the Erlang-B formula with traffic intensity $\lambda/\mu \equiv \rho$, and $N-k$ available circuits;

this assumes that if all fixed channels in a co-channel cell are busy, then an arriving call there generates an attempt on the (borrowed) channel under consideration. If on an average k_0 is taken to be N, then $V > 0$ translates to

$$\frac{N_c \lambda E_r(\rho; N - k)}{(N + 1)(\mu + \delta)} < 1. \tag{10}$$

The analysis in section 2.1, can be extended to show that if $V > 0$, then expected calls carried with dynamic allocation is better than fixed allocation. The formal proof is omitted here. (The idea is to account each cell on a own channel of cell as generating a unit revenue flow, and each borrowed call as having a revenue flow of V.)

Conditions (9) and (10) give estimates of when channel borrowing methods considered is better than fixed allocation. ((10) gives the conditions directly in terms of the model parameters.) The technique used in arriving at these estimates is to account the benefits and penalties of borrowing, and then to incorporate this in the analysis of a single cluster outlined in section 2.1. The same technique could be used to analyze other dynamic methods as well. The next section presents simulation results.

3   Simulation

The system used in the simulation study consists of 49 cells as shown in Fig. 3. I and J describe the location of cells in nonorthogonal coordinates. The number of cells per cluster is chosen to be seven. A wrapping structure is used, which connects row $J = 7$ to row $J = 1$ and column $I = 7$ to column $I = 1$, so as not to allow any traffic to leave the system.

The call arrivals in each cell are assumed Poisson, and the rate is same for all cells. The conversation time is exponential with average of 98 seconds. Both no-handoff and handoff cases are considered. In the latter, 10% of traffic on average move across cell boundaries, and the call residing time in a cell is exponential with average 88.2 seconds. Handoffs are uniform in all directions. The number of channels per cell is 10. The figure of 5 Erlangs (which yields 0.018 blocking probability for fixed assignment) was used as the starting point of the simulation. The borrowing method simulated is as described in section 2.2. For borrowing, the neighboring cells are scanned clockwise, starting from the north neighbor. A channel is borrowed from a neighbor with maximum B-set availability. In case of a tie, the one which appears first in the selection order is selected. One of the borrowable channels in the cell selected is used to carry a new call. During the duration of the call, that particular channel in the co-channel cells is locked or marked as unavailable. Normal calls which

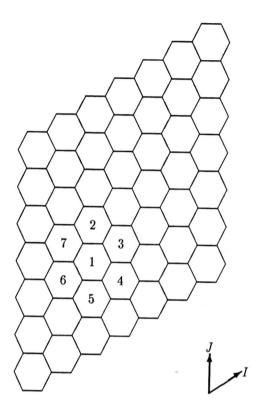

Figure 3: *The Cell Structure Used in Simulation*

use fixed channels in one cell are terminated as soon as the holding time of a channel expires in the normal pattern. But calls which use borrowable channels in their cells or borrowed channels from the neighbor cells switch to a fixed channel as soon as one is available. In the latter, in addition to the switching to a fixed channel, the channel is unlocked from the relevant cells. Calls are terminated as soon as the holding time expires, with unlocking of the channels if they used borrowed channels. Assignment procedures are same for new and handoff calls. A blocked handoff call is a forced termination.

3.1 Simulation Results

The results of simulations are shown in Fig. 4 – Fig. 7. These present the probability of first attempts as a function of the traffic intensity $\rho = \lambda/\mu$ (erlangs/channel/cell coverage area). Fig. 7 presents the probability of forced terminations. The dynamic methods perform better than fixed allocation in the low traffic region as expected (see equations (9) and (10), in terms of

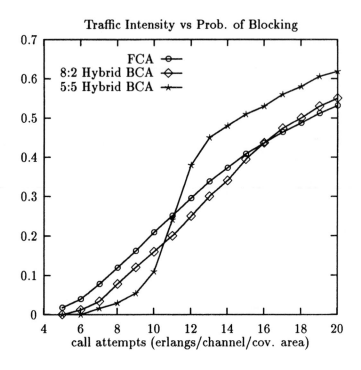

Figure 4: *Traffic Intensity vs. Blocking Probability of First Attempts (no-handoffs)*

both blocking probability of first attempts and probability of forced termination. Table 1 presents the cutoff values at which fixed and dynamic allocation curves crossover. There is close agreement between theoretical (from (10)) and simulation values. However, it should be noted that approximations were made in deriving (10); further study is needed to ascertain whether this is a good estimate for all parameter values of the model.

4 Concluding Remarks

In this paper, we have presented an analysis of channel allocation methods, using the notion of stochastic dominance for comparison of fixed and dynamic methods. Estimates of the parameters for which dynamic methods are better than fixed allocation were presented. Some approximations were made in the

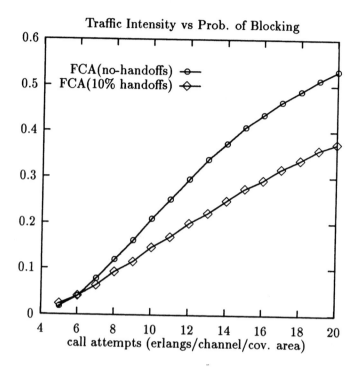

Figure 5: *Traffic Intensity vs. Blocking Probability for FCA (10% handoffs and no-handoffs)*

analysis in section 2.2. This could be made more accurate by incorporating the effects of interference and specific channel allocation procedures features more precisely (particularly for P and k_0 in equation (9)). Issues of incorporating into the model irregular cell geometries, specific propagation models (multipath, fading, shadowing), dynamic allocation of base station and channels based on interference measurements, and spatially nonuniform traffic etc. are important topics for future research.

References

[1] D. Cox and D. Reudink, "Increasing channel occupancy in large-scale mobile radio systems: dynamic channel reassignment," *IEEE Trans. on Veh. Tech.*, vol. VT-22, pp. 218-222, Nov. 1973.

140

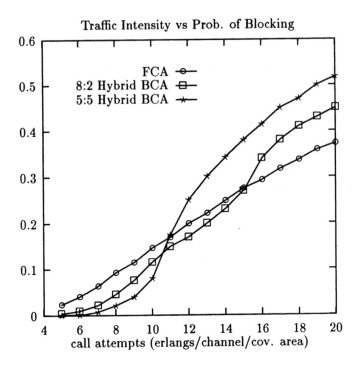

Figure 6: *Traffic Intensity vs. Blocking Probability of First Attempts (10% handoffs)*

[2] L. Schiff, "Traffic capacity of three-types of common user mobile radio communication systems," *IEEE Trans. on Commun. Tech.*, vol. COM-18, pp. 12-21, Feb. 1970.

[3] D. Everitt and D. Manfield, "Performance analysis of cellular mobile communication systems with dynamic channel assignment," *IEEE J. Select. Areas Commun.*, vol. 7, pp. 1172-1180, Oct. 1989.

[4] J. Xu and P. Mirchandani, "Channel allocation in cellular networks," University of Arizona, 1989.

[5] G. Foschni and B. Gopinath, "Sharing memory optimally," *IEEE Trans. on Commun.*, vol. 31, pp. 352-360, Mar. 1983.

[6] J. Virtamo, "Reciprocity of blocking probabilities in multiservice loss systems," *IEEE Trans. on Commun.*, vol. 36, pp. 1174-1175, Oct.1988.

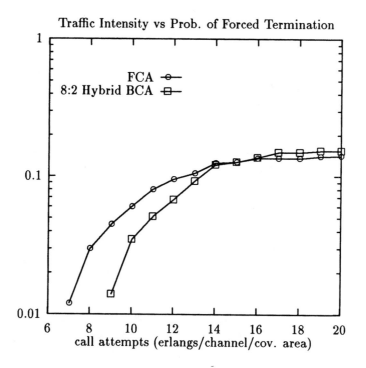

Figure 7: *Traffic Intensity vs. Probability of Forced Termination (10% hand-offs)*

[7] F. Kelly, "Blocking probabilities in large circuit-switched networks," *Advances in Applied Prob.*, vol. 18, pp. 473-505, 1986.

[8] W. Whitt, "Blocking when service is required from several facilities simultaneously," *AT&T Tech. J.*, vol. 64, pp. 1807-1856, 1985.

[9] S. Jordan and P. Varaiya, "Throughput in multiple service, multiple resource communication networks," to appear in *IEEE Trans. on Commun.*, 1990.

[10] G. Choudhury and S. Rappaport, "Cellular communication schemes using generalized fixed channel assignment and collision type request channels," *IEEE Trans. on Veh. Tech.*, vol. VT-31, pp. 53-65, May 1982.

with FCA	no-handoffs		10% handoffs	
	8:2 BCA	5:5 BCA	8:2 BCA	5:5 BCA
Theoretical (Erlangs)	15.92	10.55	15.11	9.96
Simulation (Erlangs)	15.80	11.00	15.20	10.60

Table 1: *Cut-off points for blocking probabilities*

[11] D. Hong and S. Rappaport, "Priority oriented channel access for cellular systems serving vehicular and portable radio telephones," *IEE Proc.*, vol. 36, pt. I, pp. 339-346, Oct. 1989.

[12] J. S. Engel and M. M. Peritsky, "Statistically-optimum dynamic server assignment in systems with interfering servers," *IEEE Trans. Veh. Tech.*, vol. VT-22, no. 4, pp. 203-209, Nov. 1973.

[13] T. J. Kahwa and N. D. Georganas, "A hybrid channel assignment scheme in large-scale, cellular-structured mobile communication systems," *IEEE Trans. Commun.*, vol. COM-26, no. 4, pp. 432-438, Apr. 1978.

[14] S. M. Elnoubi, R. Singh, and S. C. Gupta, "A new frequency channel assignment algorithm in high capacity mobile communication systems," *IEEE Trans. Veh. Tech.*, vol. VT-31, no. 3, pp. 125-131, Aug. 1982.

[15] M. Zhang and T. P. Yum, "Comparisons of channel-assignment strategies in cellular mobile telephone systems," *IEEE Veh. Tech.*, vol. 38, no. 4, pp. 211-215, Nov. 1989.

[16] L. G. Anderson, "A simulation study of some dynamic channel assignment algorithm in a high capacity mobile telecommunication system," *IEEE Trans. Veh. Tech.*, vol. VT-22, no. 4, pp. 210-217, Nov. 1973.

[17] S. Nanda and D. J. Goodman, "Dynamic resource acquisition : A scheme for carrier allocation in cellular systems," *Proc. of Third Gen. Wireless Info. Networks*, Rutgers, NJ, Oct. 1990.

[18] S. Ross, *Introduction to stochastic dynamic programming*, Academic Press, New York, NY, 1983.

A RADIO-LOCAL AREA NETWORK WITH EFFICIENT RESOURCE ALLOCATION

H. C. Tan, M. K. Gurcan & Z. Ioannou
Digital Communications Section, Department of Electrical Engineering
Imperial College of Science, Technology and Medicine
University of London, United Kingdom

Abstract

In this paper, we propose a cellular-based Radio-Local Area Network (R-LAN) which uses a novel dynamic resource distribution scheme. It is able to cope well with non-uniform spatial and time distributions of traffic in the cells, and also increases the bandwidth utilisation, thus enabling the network to support a higher throughput, and provide an improvement of more than 40% in the average message delay performance. This document will describe the architecture of the R-LAN, the dynamic resource allocation scheme and the results of computer simulation studies.

1. Introduction

Office automation started with the invention of the typewriter and telephone networks. Today, modern offices are filled with terminals and equipment used for information processing. To share expensive equipment, such as printers and fixed disks, and to facilitate information transfer, these machines are interconnected either by the use of existing telephone networks or by the addition of wire or fibre-optic Local Area Networks (LANs). In view of the expected increase in the number of terminals and the great demand for interconnecting them, the wireless office information network is suggested as a solution to avoid expensive cable installations, to provide mobility to equipment and to reduce relocation costs drastically. In fact, wireless networks are expected to either partially or completely replace existing fixed networks in the office.

Interest in wireless indoor communications have increased in recent years and several systems have been proposed in the United States of America and Europe. These include the Hewlett Packard radio link network, Motorola's indoor network, Electronic System Technology's wireless LAN [LESS87, ZHAN90], Digital European Cordless Telecommunication (DECT) system and the second-generation digital cordless telecommunications (CT2, a UK initiative) [SWAI90].

2. A Short Description of the Novel Scheme

The architecture of the proposed Radio-Local Area Network (R-LAN) is a hybrid fixed-link and radio network, based on the cellular concept. This is shown in Figure 1, for a four-storey building. The base-stations are connected by a fixed, high-capacity "backbone" link network, *e.g.* fibre-optic cables. The users, which are represented in the figure as computer terminals, are connected to the network by radio links. The major restriction is in the radio link between the base-station and the mobile users within the corresponding cell. With this in mind, we are proposing a novel and efficient resource allocation scheme for the uplink, to improve the throughput of this link.

The users in every cell originate calls randomly. Therefore over a short period of time there is a finite probability that a call may not have arrived in certain cells. Taking advantage of this non-uniform distribution of traffic over the service area (consisting of the cells), we propose a flexible and dynamic time division multiplexing method of distributing the transmission bandwidth.

At regular intervals of time, known as *Update Intervals*, the traffic demand is interrogated by the base-station of each cell, as shown in Figure 2. An interference graph is generated from the information obtained after each interrogation. The interference graph models the base stations as nodes, when the interrogation reveals that message(s) had arrived in the preceding update frame, and draws an edge between the nodes if they are within co-channel interfering range. If no message(s) had arrived in a cell in the preceding update frame then the node representing this cell does not appear in the interference graph. A typical interference graph, modelling the case where only eight, out of a total of sixteen, base-stations were found to need transmission capacity, *i.e.* interrogation had revealed that there were message arrivals in these cells, is shown in Figure 3.

Using an efficient method [TAN90], we determine the minimum number of time slots required for non-interference between the cells, and also the distribution of these time slots amongst the cells. For example, the interference graph of Figure 3 requires only three time slots, as determined by this algorithm. Furthermore, the duration of the allocated time slots corresponds to the number of packets that needs to be transmitted in each cell. As this procedure is repeated at every update, the resource allocation is dynamic and efficient.

In this way, arriving messages are queued in a buffer and transmitted only after the duration of the available time slot is allocated and assigned to them. This leads to an efficient distribution of resources only to cells that require transmission capacity. In addition, during the allocated time slots the data packets are transmitted synchronously and continuously, thus providing the following advantages :

    1. the use of simple synchronising schemes at the base-station receiver and

    2. efficient utilisation of the available slot time.

The performance of this scheme was compared to that of a Static Allocation Time Division Multiplex (SATDM) scheme where a fixed proportion of the transmission capacity was allocated to each cell. It was found that the average packet delay in the SATDM scheme was more than 50% higher than that for the dynamic scheme investigated here. This scheme is also able to support higher arrival rates. Its flexible and dynamic nature also enables it to cope extremely well when the average arrival rate of each cell is different. Consequently, it will also perform well when there are time fluctuations in the traffic demand. Time and spatial fluctuations in traffic demand is a characteristic of any cellular system and it is vital that the system is able to respond efficiently. Through proper control of the update frame and packet durations, the performance of the R-LAN can be further improved. By using overlapping interference graphs, it is also possible to support various services in the R-LAN that require different levels of bit error rates (BERs).

3. Dynamic Resource Distribution

Having obtained the interference graph, we utilise an algorithm to determine the minimum number of time slots required for the users in each cell to transmit on, without causing excessive interference to other users in the network. The algorithm we use will also determine the explicit distribution of these time slots between the various cells that require transmission capacity. Let's call this algorithm the Dynamic algorithm.

If the method is applied to the example of Figure 4, the base-stations are grouped as follows :

    1. $\{B_{02}, B_{04}\}$ - Group 1
    2. $\{B_{13}, B_{01}, B_{07}\}$ - Group 2
    3. $\{B_{05}, B_{12}, B_{14}\}$ - Group 3

The Dynamic algorithm divides the update interval into a minimum number of time slots. Hence, individual time slots can be of a longer duration. Consequently these time slots are also reused as closely as possible. This results in increased bandwidth utilisation and better network performance in terms of throughput and average packet delay.

The resource allocation scheme can be made to correspond even more closely to the demand by varying the proportions T_1/T_U, T_2/T_U and T_3/T_U in sympathy with the number of packets waiting for transmission capacity in the corresponding cells, where T_1, T_2 and T_3 indicate the durations of the time slots allocated to the cells of Groups 1, 2 and 3 respectively and T_U is the update interval. Determine, for the example of Figure 4,

$$P_{max1} = \max(P_{02}, P_{04})$$
$$P_{max2} = \max(P_{13}, P_{01}, P_{07})$$
$$P_{max3} = \max(P_{05}, P_{12}, P_{14})$$

Where the number of packets in cell B_{01}, B_{02}, ... are P_{01}, P_{02}, ... and
$$P_T = P_{max1} + P_{max2} + P_{max3}$$

The time division ratios are then calculated as follows :

$$\frac{T_1}{T_U} = \frac{P_{max1}}{P_T} \qquad \frac{T_3}{T_U} = \frac{P_{max3}}{P_T} \qquad \frac{T_2}{T_U} = \frac{P_{max2}}{P_T}$$

The reason for this will be clear when we examine the queueing model for the R-LAN, as shown in Figure 5. As far as the users in the cells are concerned, they will encounter a central buffer in each cell, where the packets will be queued. After the algorithm has determined the time slot allocations, the buffers will be serviced at the appropriate times. Individual buffers will see a single server that provides intermittent service *i.e.* available only for a specified proportion of the update interval as shown in Figure 5.

The effect of the time slot reuse, as determined by the algorithm, is to group base-stations that can utilise the same time slot together. These base-stations are then served simultaneously. Therefore, only the base-station with the maximum number of packets in each Group is critical in determining the required transmission capacity. If the time slot duration allocated is sufficient for the cell with the highest number of packets in queue, then by the end of the time slot, it is certain that the packets, which are lesser in number, in the other cells will have been served completely.

4. Characterising the Delay Performance of the R-LAN

Figure 6 shows the Average Packet Delay as a function of Update Duration. Analytical studies of the delay performance at cell level is given in [IOAN90]. Each curve is for a particular combination of Mean Interarrival Time and Packet Length. Each data point represents the steady state value of the Average Packet Delay.

Comments

1. It can be seen that the curve for Mean Interarrival Time = 0.05 seconds and Packet Length = 0.006 seconds represent the boundary of stable operation. If either the Mean Interarrival Time is further reduced or the Packet Length is further increased, or both, then the queue in the buffer builds up indefinitely and the queuing delay increases rapidly for certain arrival rates.

2. Having said this, we should also note that the curves for Mean Interarrival Time = 0.04 seconds and Packet Length = 0.006 seconds, and Mean Interarrival Time = 0.05 seconds and Packet Length = 0.007 seconds, for values of Update Duration from 0.065 to 0.105 seconds are stable. This shows that if higher packet delays can be tolerated, then higher packet arrival rates or packet length can be supported by the R-LAN.

3. It is also clear that the system is unstable in two regions :
   a. when the value of the Update Duration is small, *i.e.* < 0.025 seconds and
   b. in a band of Update Durations centred around 0.05 seconds.

This is an interesting result that needs further investigation.

Figure 7 is an important graph. It shows how the average Packet Delay varies with respect to the Packet Arrival Rate for both the Dynamic and the SATDM schemes. The spatial distribution of the traffic includes both uniform and non-uniform patterns. There are more results than can be included.

5. Conclusions and Comments

1. Figure 7 shows that Average Packet Delay for the Dynamic scheme is lower than that for the SATDM scheme over the whole range of Packet Arrival Rates. This is a clear indication that the Dynamic scheme is better and more efficient at utilising the bandwidth than SATDM. The

bandwidth available to both schemes and the traffic demand is the same. Therefore, the improvement is due entirely to the dynamic and efficient reuse of the bandwidth afforded by the scheme.

2. The mean value of the reduction in Average Packet Delay is 0.015 seconds, representing an improvement of 47.4%.

3. The end of each curve on the right hand side represent the maximum Packet Arrival Rate that the network can cope with before going into instability.

4. For both 2-cell and 4-cell non-uniform traffic distributions, the Dynamic scheme provides superior performance. The improvement is reflected in terms of :
    a. Increased throughput. For the 4-cell pattern, the Dynamic scheme can support up to 20.83 packets per second, while SATDM can support up to 19.23 packets per second. The figures for the 2-cell pattern are 25 and 17.31 packets per second correspondingly.
    b. Reduced Average Packet Delay over the whole range of operation. The reductions for the 4-cell and the 2-cell patterns are 66.43% and 60.02% respectively.

5. From the results above, it can be seen that the improvements in packet delay, due to the use of the Dynamic scheme, becomes more pronounced as the traffic becomes more non-uniformly distributed, as in the 4-cell pattern. However, the throughput for the 2-cell pattern is higher than that for the 4-cell pattern.

6. For certain combinations of the system parameters : Packet Length, Update Duration and Mean Interarrival Time, the available capacity is more than sufficient to cope with the traffic demand. In these cases, the Average Packet Delay is dependent on the Update Duration. The Average Packet Delay in stable regions can be predicted as follows

$$T_D = \tfrac{5}{6} T_U + T_S$$

where T_D is the average packet delay, T_U is the update interval and T_S is the packet transmission time.

7. For some other combinations of the system parameters mentioned above, the R-LAN becomes unstable. The Average Packet Delay builds up indefinitely in these cases. We have found that the R-LAN can

support traffic with a minimum Mean Interarrival Time of 0.05 seconds and a maximum Packet Length of 0.006 seconds, over the whole range of Update Durations from 0.015 to 0.105 seconds (Figure 6). This defines the stable operating range of the R-LAN. The unstable regions of the R-LAN require further investigation.

REFERENCES

[IOAN90]   Ioannou, Z., "Multiplexing and Multiple Access Techniques for Radio-Local Area Networks," *M.Phil/Ph.D. Transfer Thesis*, Elect. Engg. Dept., Imperial College London, United Kingdom, August 1990. Also paper to appear in this workshop.

[LESS87]   Lessard, A., "Wireless Communications for an Automated Factory Environment," *Ph.D. Dissertation*, Academic, Computer Science, UCLA, 1987.

[SWAI90]   Swain, R. S., "Digital Cordless Telecommunications - CT2," *British Telecommunications Engineering*, Vol. 9, July 1990.

[TAN90]   Tan, H. C., "Dynamic Resource Allocation for Radio-Local Area Networks," *M.Phil/Ph.D. Transfer Thesis*, Elect. Engg. Dept., Imperial College London, United Kingdom, August 1990.

[ZHAN90]   Zhang, K., Pahlavan, K., "An Integrated Voice/Data System for Mobile Indoor Radio Networks,", *IEEE Trans. Veh. Tech.*, Vol. 39, No. 1, pp. 75 - 82, Feb. 1990.

150

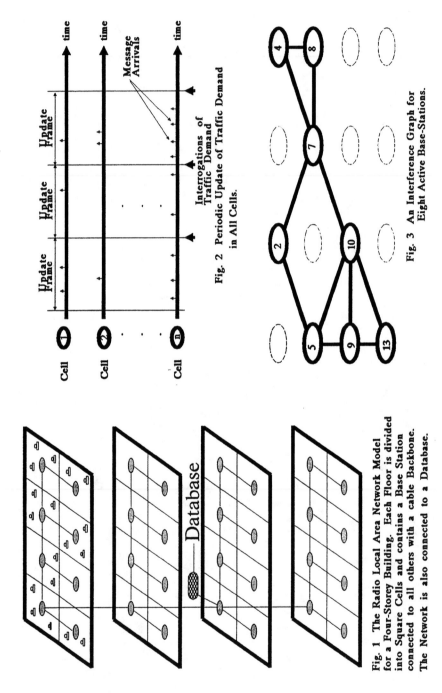

Fig. 2 Periodic Update of Traffic Demand in All Cells.

Fig. 3 An Interference Graph for Eight Active Base-Stations.

Fig. 1 The Radio Local Area Network Model for a Four-Storey Building. Each Floor is divided into Square Cells and contains a Base Station connected to all others with a cable Backbone. The Network is also connected to a Database.

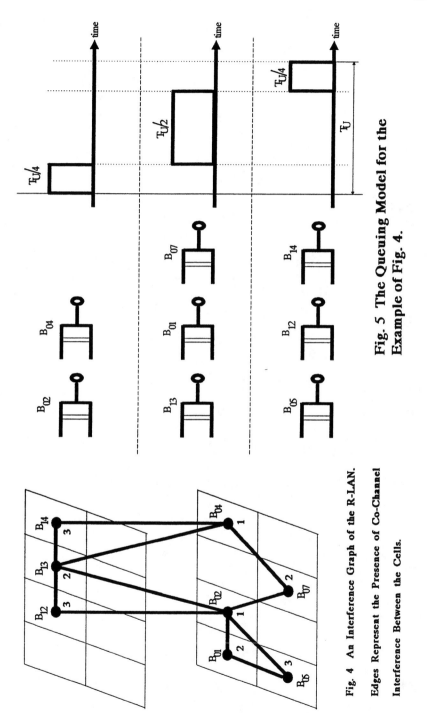

Fig. 5 The Queuing Model for the Example of Fig. 4.

Fig. 4 An Interference Graph of the R-LAN.

Edges Represent the Presence of Co-Channel Interference Between the Cells.

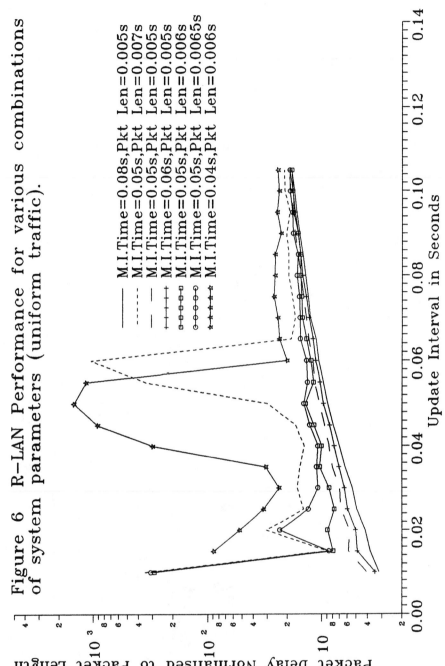

Figure 6 R–LAN Performance for various combinations of system parameters (uniform traffic).

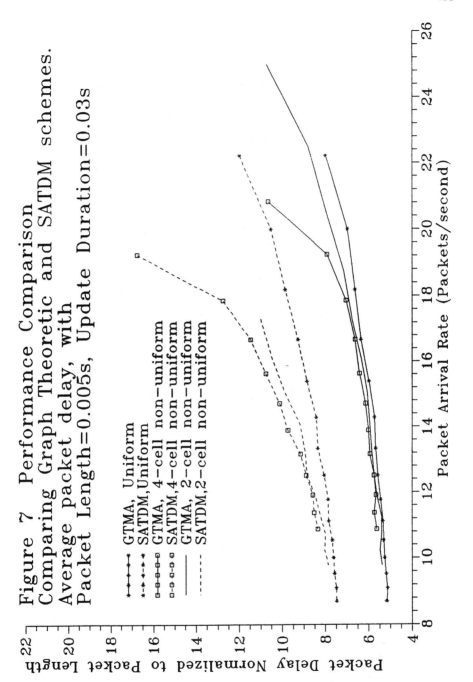

Figure 7 Performance Comparison
Comparing Graph Theoretic and SATDM schemes.
Average packet delay, with
Packet Length=0.005s, Update Duration=0.03s

Radio Channel Control for A Multi-Carrier TDMA Microcell System

Izumi Horikawa, Masahiko Hirono, and Kazushige Tanaka

Mobile Communications Division

Nippon Telegraph and Telephone Corporation

Abstract

This paper discusses a radio control channel architecture for a microcellular system which has a radio interface which provides transmission bit rate of less than 500 kb/s, short TDMA frame length of multi-carrier TDM systems, and then presents a physical channel structure allocating broadcasting and common control channels on one TDM carrier frequency suited for decentralized and dynamic channel control of the system.

1. Introduction

A radio channel architecture that provides accommodation for high density traffic, voice and nonvoice teleservices, and low-cost pocket wireless unit is important for future microcell systems.

As has been reported in many previous papers, dynamic channel assignment, DCA, is a distinctive requirement for effective use of radio spectrum in the irregular microcell pattern. High speed hand-over is also important in microcellular systems since it enables them to provide seamless services.

A multi-carrier TDMA scheme in which TDMA carriers with several channels are multiplexed in frequency domain is necessary to provide the more than one hundred radio channels required to handle such heavy traffic loads as 20,000 E/km^2.

Some strategies to enable quick selection of the best channel and the alternatives from a frequency-time matrix are required in the multi-carrier TDMA system at both mobile and base stations for DCA and hand-over.

A new radio interface with low delay frame length for the next generation microcellular systems is expected to be used in a wide range of service applications such as residential cordless telephones, wireless PBXs, and personal communications networks, allowing speech quality equivalent to that of fixed telephones without the need for sophisticated echo control devices.

This paper reviews the characteristics of a radio channel structure and control architecture taking TDMA frame length of multi-carrier TDM systems into consideration, and then presents a physical channel structure allocating broadcasting and common control channels on one TDM carrier frequency suited for decentralized and dynamic control of the microcelullar systems.

2. Background

Low Delay Time Frame Structure

The speech quality of the next generation wireless telephone systems should be as near to that of fixed telephones as possible.

A CODEC algorithm mainly determines speech quality and a 32 kb/s ADPCM is suited for the next generation wireless telephone systems from the view points of both quality and delay time.

A shorter delay time frame structure as well as a CODEC which does not need any sophisticated echo control devices is important to implement a portable unit with low cost for the residential as well as the business environment.

Speech quality is mainly degraded by near-end echo, although there are two different path echoes in wireless access systems, as shown in Fig. 1.

A frame length of around 5 ms is recommended, taking the relative signaling and guard time overhead in a radio slot into consideration.

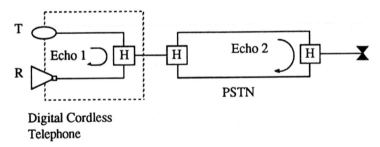

Digital Cordless
Telephone

Fig. 1 Simplified Echo Path Model

Common Control Channel Scheme

There are two common control channel schemes, dedicated and associated.

The dedicated scheme, in which a physical channel is dedicated in each cell for common control, has been utilized in analog cellular systems and will be utilized in digital cellular systems.

The associated scheme is one, in which a control signal, a so called "beacon," is provided which is associated with busy traffic channels in the fix-to-mobile direction, and with an idle traffic channel in the mobile-to-fix direction, as in the DECT system [1].

Some problems are pointed out in the latter scheme. One of these is that, in time frame segmentation, control signals, including BS-ID and signals for both common and in-service control, are contained in all frequency-time channels. This results in excess capacity for control and deteriorates the efficiency of overall radio spectrum utilization.

Table 1 Comparative Time Segmentation

System	GSM	USA DC	Japan DC	DECT	This Paper
CC CH	Dedicated	Dedicated	Dedicated	Associated	Dedicated
OH/Total	0.26	0.25	0.19	0.19	0.27
Signal/Total	-----	0.09	0.11	0.11	0.18
Burst Length [ms]	0.6	6.67	6.67	0.42	0.63
Channel Spacing (kHz)	200	30	25	1728	500

Another problem is that several frames are required for one-signal transmission. This is because the number of bits allowed for the control data including the FEC is limited to about 60 to maintain a slot efficiency of about 70-80%, as shown in Table 1. This results in less effective battery saving performance, since the PS has to be remain awake longer to receive a complete set of control signals in the sleeping mode.

Multi-Carrier TDMA

The maximum radio transmission bit rate is limited by waveform distortion caused by multipath delay propagation. Given the bit rate and frame length, the TDM channel number is determined.

DECT will be operated at a bit rate of 1.1 Mb/s with GMSK. Under these conditions, however, severe BER degradation has been seen in our experimental investigations of 256 kb/s transmission in a modern office surrounded by metal walls. The transmission bit rate should be carefully chosen so that sufficiently low BER can be obtained without an equalizer.

In any event, a multi-carrier TDMA scheme in which TDMA carriers with several channels are multiplexed in the frequency domain is necessary to provide the more than one hundred radio channels required to handle such heavy traffic loads as 20,000 E/km^2 [2].

High Speed Carrier Switching

By switching the carrier frequency slot by slot during the talking period, a portable unit can easily monitor the surrounding base stations' control channel, since planned channel allocation is applied and a control channel is designated for each base station in the cellular system.

In a microcellular system, the control channel as well as the traffic channel should be allocated on a DCA basis. To complete the monitoring in microcellular systems, the portable unit has to scan the entire frequency-time channel matrix, switching the carrier frequency within a guard time of between 30 and 50 microseconds.

The wider the carrier spacing of the system, the shorter the carrier switching time of a PLL frequency synthesizer becomes. In the case of DECT, channel spacing of 1.728 MHz seems to provide a satisfactorily short carrier switching time in the control architecture.

Figure 2 shows the calculated acquisition time for the various frequency separations for a frequency synthesizer with preset frequency control [4], which is expected to provide a shorter acquisition time than that of a conventional unit.

Figure 2 also shows, however, that even if the new PLL synthesizer is adopted to the narrow carrier spacing system, it can not complete the switching within the guard time. Other frequency synthesizers, such as the multi-oscillator switching type and the direct digital type, are superior from the stand point of acquisition time, but are larger in size and more power consumptive than the PLL synthesizer.

In the following section, a radio channel structure which has potential decentralized and mobile-controlled radio access and Hand-over is proposed, taking into considerations of short delay time frame of about 5 ms, radio channel separation of less than 500 kHz, and high traffic density of 20,000 E/km^2 for indoor application.

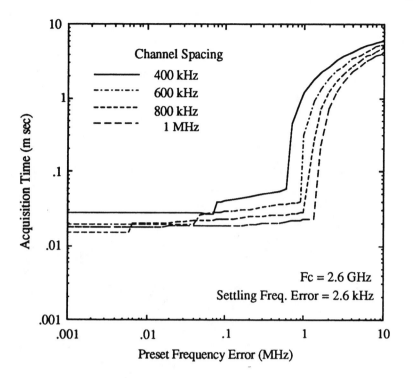

Fig. 2 Acquisition Time of PLL Synthesizer with Preset Frequency Control

3. Channel Structure

Dedicated Common Control Channel

Microcellular systems require a channel structure which makes the PSs sensitive to the nearby BSs during both sleeping and communicating time, so that mobile controlled radio access and hand-over can be realized in such systems.

By concentrating broadcasting and common control channels for each BS on TDM slot(s), the PSs can quickly scan the BSs to compare the received power levels from the surrounding BSs. The TDM slots with super frames are exclusively allocated to each BS on a dynamic channel assignment basis, as is shown in Fig. 4 [2], [3].

The traffic channels for each BS can be selected at any different carrier frequency from the control carrier frequency with a carrier-switching TDMA transceiver. This scheme combines the advantages of FDMA and TDMA with efficient radio channel utilization in high interference environments, as shown in Fig. 3.

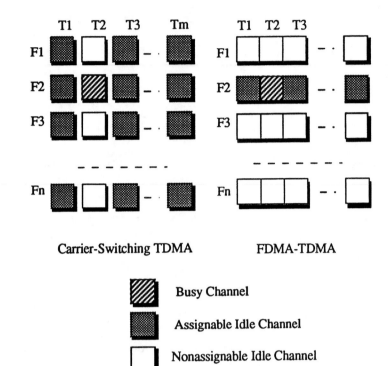

Fig. 3 Assignable Channel Area Comparison of Multi-Carrier TDMA

With a designated control carrier scheme, site diversity transmission for down link control signals and site diversity reception on the uplink control signals will perform wide area paging and secure handshake in the call set-up.

This scheme also provides the feature of matched control channel capacity for microcell traffic.

Furthermore, the BSID and all the control data can be transmitted on one burst using the Data and I areas of the radio packets, making this an effective operation under a interference limited environment.

Radio Zone Access and Traffic Channel Assignment

Dynamic channel assignment is a distinctive requirement for effective use of a radio spectrum in the irregular microcell pattern, as has been reported in many previous papers. In selecting the final channel candidate for this assignment, the no-threshold strategy based on D/U appears to be more efficient in terms of spectrum utilization than the fixed threshold strategy.

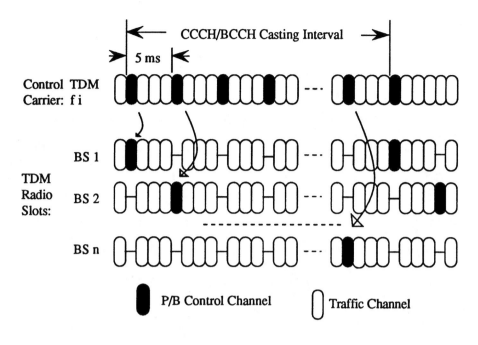

Fig. 4 Radio Control Channel Structure:
Sequential Slot Sharing of One TDM Carrier

The BSs store some channels with the lowest undesired power level in the memory for subsequent call set-up, periodically supervising the undesired power level on the idle traffic channels.

PSs, when sleeping, intermittently monitor the designated control carrier slot to select the base station with the maximum D/U.

The traffic channel assignment is conducted by a BS assisted by the responding mobile after the BS receives a call set-up message, regardless of whether the call request is from a mobile or fixed network. After this, the D and U levels on the traffic channel at the portable side are measured. When both the PS and BS have approved the traffic channel allowable for the undesired level and for D/U, then the channel can be used for communication.

In this manner, the dynamic channel assignment by the BS assisted by the mobile will be effectively achieved.

Hand-over

Mobile controlled hand-over can be also introduced into the proposed control architecture.

When the PS detects BER degradation, it starts hunting for a new BS, since it can not switch the carrier frequency to monitor the BSs during talk periods. While scanning the BSs, the PS can continue talking on the traffic channel on the control carrier frequency if the old BS is still within the range.

When the PS can not communicate with the old BS, as sometimes happens when shadowing causes fast CNR degradation, the PS quickly sets up a radio channel with a new BS by the call origination procedure. This process is completed within a few several hundred milliseconds, since the super frame of the control channel to be monitored is constructed within that period. The fast CNR degradation which seems to frequently occur in micro/pico-cellular systems is effectively managed by this control architecture.

When CIR degradation causes low power hand-over to occur, the PS first tries to find a new channel within the current BS. By taking advantage of the multiple channels in a TDMA transceiver, this can be completed with hitless switching. The other types of low power hand-over, which prevents excess cochannel interference, can only be initiated by a PS with an agile synthesizer which searches for better BSs during communications.

4. Conclusion

This paper has reported a physical channel structure suited for such microcellular systems with a low delay time frame, channel spacing of less than about 500 kb/s, and a multi-carrier TDMA.

A dedicated common control channel scheme, allocating broadcasting and common control channels on one TDM carrier frequency, is proposed for providing mobile controlled radio zone access and hand-over.

The carrier-switching TDMA effectively introduces this control channel structure, and provides dynamic and decentralized channel control.

References

[1] Heinz Orchsner,"Radio Aspects of DECT," The 4th Nordic Seminar, Oslo, 1990.

[2] Izumi Horikawa and Masaaki Hirono,"Multi-Carrier Switching TDMA-TDD Microcell Telecommunications System," IEEE VTC'90, pp. 167-171, Orland, 1990.

[3] Izumi Horikawa and Masaaki Hirono,"A Digital FDMA/TDMA Micro cell System for the Next Generation of Cordless Telephones,"The 4th Nordic Seminar, Oslo, 1990.

[4] Yoshiaki Tarusawa and Yasushi Yamao "Hi-Speed Digital Loop-Preset (DLP) Frequency Synthesizer," 1989 Spring National Convention Record, IEICE of Japan, No. B-545, 1989.

MODELS FOR CALL HAND-OFF SCHEMES IN CELLULAR COMMUNICATION NETWORKS

Stephen S. Rappaport

Department of Electrical Engineering
State University of New York
Stony Brook, New York 11794-2350

ABSTRACT

In the past much work that considered the problem of hand-off in cellular communications systems, relied exclusively on simulation models. However, it is possible to create reasonable and tractable analytical models which provide additional insight and permit computation of theoretical performance characteristics. This paper presents an underlying methodology that can be used to model a variety of issues and schemes relating to the problem of hand-off. The approach, which is based on the notion of multi-dimensional birth-death processes, is rich enough to permit representation of many factors, including (among others): a broad class of probability laws that describe the underlying driving processes such as session holding times and platform mobility factors; platforms that support multiple calls; various priority oriented schemes; and, both large and small user populations. Some example applications and performance results are given.

The research reported here was supported in part by IST/SDIO and administered by the U.S. Office of Naval Research under Contract no. N00014-85-K0610.

Figures are included at the end of the paper.

INTRODUCTION

The rapid growth of cellular and wireless communications networks presents an array of technical challenges which must be met in order to provide high quality service to users. Among these is the problem of hand-off in which calls on board communicating mobile platforms that cross spatial zones must be allocated communications resources in the target zone. This arises in systems that use circuit-switched or virtual circuit-switched protocols and for which the likelihood is high that more than one wireless gateway is needed to provide continuous service to a mobile platform during a single call session. Hand-off attempts which fail result in <u>forced termination</u> of calls and are perceived by users to be interruptions of service. The phenomenon is considerably more obtrusive than simple call <u>blocking</u>, for which a call attempt is <u>initially</u> denied access because of channel unavailability. Some priority for hand-off attempts may be desirable.

The hand-off problem is particularly important in high capacity cellular communications in which increased spectral efficiency is achieved by using small zones, sectors, microcells, etc. While this approach allows greater spectral reuse over the region served by the system, it also requires more transceiver sites, and in operation tends to generate more cell boundary crossings by mobiles with calls in progress.

Early studies of the problem relied exclusively on computer simulation [1]. More recent work however, attempts to develop realistic analytical models [2-6]. These can provide greater insight, and are computationally more efficient in that they allow investigation over a larger set of parameters.

While, the hand-off problem is becoming increasingly significant for terrestrial cellular communications as mobile and portable communications proliferate, it also arises in other mobile communications scenarios.

Figures 1-3 suggest various settings. Generally, when circuit switched or virtual circuit switched protocols are used with networked gateways equipped for tetherless communications, hand-off problems arise when a communicating mobile platform moves out of range of its current gateway.

FEATURES OF THE CALL HAND-OFF PROBLEM

The following features characterize the hand-off problem in a general way. There is a spatial region that is served by networked gateways that provide links to mobile platforms. Often the region is tessellated by cells / satellite beams / volumes / zones / sectors / microcells / picocells. We use the term cells in its generic sense to describe any or all of these. Large numbers of mobile platforms traverse the region. At each gateway (or cell) there may be constraints on channel or spectral use, and on platform service. Examples of the former include a limit on the total number of channels that can be used at a gateway at any single time, the number of modems deployed at a gateway, and channel use quotas for different platform types or classes. Similarly, constraints on platform service may include an upper limit on the number of platforms that that can be served simultaneously, and specific platform quotas based on platform type or class. Additionally, we can consider the single call hand-off problem, in which each mobile platform can support at most one call at a time, and the multiple call hand-off problem, in which at least some platforms can support a multiplicity of calls. Examples of the former are persons, private automobiles, and single-passenger taxicabs. Other types of mobile platforms, such as buses, planes, trains, and ferry-boats can support multiple calls simultaneously. The essential feature in the multiple-call hand-off problem is that a cell boundary crossing by a single platform can generate multiple hand-off attempts.

ANALYTICAL MODELING OF CALL HAND-OFF PROBLEMS

We have been developing a methodology for modeling call hand-off problems. The approach which makes use of the notion of multi-dimensional birth-death processes, is rich enough to include many issues involved, including: single and multiple call hand-off problems; finite and infinite user populations; priority oriented schemes for hand-off calls, -- such as cut-off priority, waiting lines for hand-off attempts while platforms are in a hand-off region, and priorities based on user platform classes and/or mobility --; a broad class of holding time distributions; a broad class of dwell time distributions; and a multiplicity of platform types that are distinguishable by mobility characteristics, number of simultaneous calls that can be supported, call generation rates, session durations, and priority class.

The basic steps to be taken in developing an analytical model for a particular situation are to:
1. Express the problem mathematically.
2. Formulate an inter-related set of characterizations and "reasonable" assumptions that render the problem amenable to an analytical or numerical solution.
   * Characterize the state of the system.
   * Recognize and characterize the underlying driving processes.
   * Characterize state transitions.
   * Develop an algorithm to solve the equations.
3. Determine performance characteristics.

We proceed by way of an example.

EXAMPLE PROBLEM STATEMENT -
MULTIPLE CALL HAND-OFF PROBLEM

There are G types of mobile platforms, indexed by $g=1,2,...G$.

A platform of type g can support up to $N(g)$ calls.

CHANNEL LIMIT.
Each cell or gateway can accommodate C channels.

CUT-OFF PRIORITY.
C_h channels in each cell are reserved for hand-off calls. New calls will be blocked if the number of channels in use is $C-C_h$ or greater. Hand-off attempts will fail if the number of channels in use is C.

CHANNEL QUOTAS.
At any gateway, the maximum number of channels that can be simultaneously used by g-type platforms is $J(g)$.

PLATFORM LIMIT.
The maximum number of communicating platforms at a gateway is V_{max}.

PLATFORM QUOTAS.
The maximum number of communicating g-type platforms at a gateway is $W(g)$.

The new call origination rate per idle port on a g-type platform with i calls in progress is $\Lambda_n(g,i)$.

The unencumbered call duration on a g-type platform is a random variable, $T(g)$, having a mean of $1/\mu(g)$.

The dwell time in a cell for a g-type platform is a random variable, $T_d(g)$ having a mean of $1/\mu_d(g)$.

Find: Blocking and Hand-off Failure Probabilities.

Note: We consider blocking probability to be the average fraction of new call originations that are denied access to a channel. Hand-off failure probability is the average fraction of hand-off attempts that fail to gain access to a channel in the target zone.

STATE CHARACTERIZATION

First consider a _single_ cell. We define the state by a sequence of non-negative integers. These can be conveniently written as G n-tuples.

$$v_{11}, \ v_{12}, \ v_{13}, \ \cdots \ v_{1N(1)}$$
$$v_{21}, \ v_{22}, \ v_{23}, \ \cdots\cdots\cdots\cdots v_{2N(2)}$$
$$\vdots \qquad \vdots \qquad \vdots \qquad\qquad\qquad \vdots$$
$$\vdots \qquad \vdots \qquad \vdots \qquad\qquad\qquad \vdots$$
$$v_{g1}, \ v_{g2}, \ v_{g3}, \ \cdots\cdots\cdots\cdots\cdots v_{gN(g)}$$
$$\vdots \qquad \vdots \qquad \vdots \qquad\qquad\qquad\qquad \vdots$$
$$v_{G1}, \ v_{G2}, \ v_{G3}, \ \cdots\cdots\cdots\cdots\cdots\cdots v_{GN(G)}$$

where v_{gi} { $g=1,2,\ldots G$; $i=1,2,\ldots N(g)$ } is the number of platforms of type g that have exactly i calls in progress. It was found convenient to order the states using an index $s=0,1,2,\ldots s_{max}$. Then the state variables v_{gi}, can be shown explicitly dependent on the state. That is, $v_{gi} = v(s,g,i)$.

When the cell (gateway) is in state, s, the following characteristics can be determined:

The number of communicating g-type platforms is

$$w(s,g) \ = \ \sum_{i=1}^{N(g)} v(s,g,i)$$

The total number of communicating platforms is

$$w(s) \ = \ \sum_{g=1}^{G} w(s,g)$$

The number of channels being used by g-type platforms is

$$j(s,g) \ = \ \sum_{i=1}^{N(g)} i \cdot v(s,g,i)$$

The total number of channels in use is

$$j(s) = \sum_{g=1}^{G} j(s,g)$$

Permissible states correspond to those sequences for which all constraints are met.

CHANNEL LIMIT	$j(s) \leq C$
CHANNEL QUOTAS	$j(s,g) \leq J(g)$
PLATFORM LIMIT	$w(s) \leq V_{max}$
PLATFORM QUOTAS	$w(s,g) \leq W(g)$

Additional constraints can also be considered within this same framework.

An ordered list of permissible states for the case of 1 platform type, up to 3 calls per platform, and 8 channels per cell (gateway) is shown in Table 1. In this case there are 41 states for a cell. Table 2 shows the number of states per cell for other parameter choices. The parameter choices above correspond to case 5 in Table 2 with C=8. As another example, consider case 3 in Table 2, which might represent sytems with G=3 platform types (say, persons, automobiles, and buses), in which persons and automobiles can support up to 1 call each, and buses can support up to 3 calls each. Thus, $N(1)=1$, $N(2)=1$, and $N(3)=3$. In addition to the channel limit, C, this case also includes a platform constraint so that no gateway will support more than 4 buses at any time, i.e., $W(3)=4$.

Even for the modest parameters of case 5 with C=8, there are 41 cell states. Since state transitions of adjacent cells are related because of hand-offs, an overall <u>system</u> state is characterized by the state for each <u>cell</u>. A system with only 7 cells would have

$(41)^7 = 1.95 \cdot 10^{11}$ states.

Clearly some additional reasonable assumptions are needed!

TABLE 1

PERMISSIBLE CELL STATES
$G = 1$, $N(1) = 3$, $C = 8$, $V_{max} \geq 8$

STATES

s	j	w	v_{11}	v_{12}	v_{13}
0	0	0	0	0	0
1	I	1	1	0	0
2	2	1	0	1	0
3	2	2	2	0	0
4	3	1	0	0	1
5	3	2	1	1	0
6	3	3	3	0	0
7	4	2	0	2	0
8	4	2	1	0	1
9	4	3	2	1	0
10	4	4	4	0	0
11	5	2	0	1	1
12	5	3	1	2	0
13	5	3	2	0	1
14	5	4	3	1	0
15	5	5	5	0	0
16	6	2	0	0	2
17	6	3	0	3	0
18	6	3	1	1	1
19	6	4	2	2	0
20	6	4	3	0	1
21	6	5	4	1	0
22	6	6	6	0	0
23	7	3	0	2	1
24	7	3	1	0	2
25	7	4	1	3	0
26	7	4	2	1	1
27	7	5	3	2	0
28	7	5	4	0	1
29	7	6	5	1	0
30	7	7	7	0	0
31	8	3	0	1	2
32	8	4	0	4	0
33	8	4	1	2	1
34	8	4	2	0	2
35	8	5	2	3	0
36	8	5	3	1	1
37	8	6	4	2	0
38	8	6	5	0	1
39	8	7	6	1	0
40	8	8	8	0	0

THIS EXAMPLE: TOTAL OF 41 STATES.

TABLE 2

**NUMBER OF CELL STATES REQUIRED FOR
MULTI-DIMENSIONAL BIRTH-DEATH PROCESS MODELS**

C	Case 1	Case 2	Case 3	Case 4	Case 5
5	91	80	90	41	16
6	155	130	150	64	23
7	250	200	234	95	31
8	386	295	346	136	41
9	575	420	489	189	53
10	831	581	665	256	67
12	1611	1036	1120	441	102
14	2886	1716	1715	711	147
16	4860	2685	2450	1089	204
18	7786	4015	3325	1600	274
20	11972	5786	4340	2272	358
22	17786	8086	5495	3136	458
24	25661	11011	6790	4225	575
26	36101	14665	8225	5575	710
28	49686	19160	9800	7225	865
30	67077	24616	11515	9216	1041

Case	G	N(1),N(2),...N(G)	Constraints
1	3	1,1,3	channel limit = C
2	3	1,1,2	channel limit = C
3	3	1,1,3	channel limit = C platform quota W(3)=4
4	2	1,3	channel limit = C
5	1	3	channel limit = C

ASSUMPTIONS

It is assumed that the system is <u>homogeneous</u>. That is, the parameters of the underlying processes are the same in any cell and all cells are statistically identical. Furthermore, we consider the system to be in <u>statistical equilibrium</u>. That is, the <u>probabilities</u> and parameters are not changing with time. Note that there are random variations with time but the <u>probability laws</u> that describe these variations do not. The system has settled down to steady state random behavior.

It is then noted that, because any platform leaving a cell enters another cell, any hand-off arrival corresponds to a hand-off departure from some other cell. In view of the homogeneity and equilibrium assumptions, it follows that,

the average hand-off ARRIVAL rate to a cell
equals
the average hand-off DEPARTURE rate from a cell.

Thus some driving processes are coupled in an "average" sense rather than in the "direct" sense that led to an unmanageable number of states. Then we consider a single given cell, and, while realizing that it is <u>typical</u>, we further assume that neighboring cells exhibit the same <u>typical</u> behavior independently. The result is that we only have to focus on a <u>single cell</u> and deal with the number of states needed to characterize its behavior. While Table 2 shows that even this can be formidable for certain parameter choices, it is at least, less dreadful!

THE DRIVING PROCESSES

The underlying processes that drive the system are as follows: {n} The generation of *new* calls in the cell of interest; {c} The *completion* of calls in the cell of interest; {h} The arrival of communicating vehicles at the cell of interest. (The cell of interest is the target cell in which channel demands are made to accommodate these *hand-off* calls.); and, {d} The *departure* of communicating vehicles from the cell of interest. (The cell of interest is the source cell from which these communicating vehicles leave.) For now we ignore the coupling between {h} and {d}. This will be accounted for in the solution algorithm. It is also noted that all of these processes are multi-dimensional. For example, a call can originate or be completed on a g-type platform, that has i calls in progress. Similarly, hand-off arrivals and departures are characterized by the platform type, g, and by the order, k, of the hand-off, where k= 1,2,...N(g).

Markovian assumptions for the driving processes, in addition to those previously discussed, render the problem amenable to solution using multi-dimensional birth-death processes. Thus it is assumed that: {1} The new call arrival processes in any state are Poisson point processes with state dependent means; {2} The unencumbered session duration, $T(g)$, for a call on a g-type platform, is a random variable having a negative exponential probability density function; {3} The hand-off call arrival process in any state follows a Poisson point process; {4} The dwell time of a g-type platform in a cell, $T_D(g)$, is a random variable having a negative exponential probability density function. Many generalizations can be handled within the same framework. For example, the unencumbered session duration for a g-type platform can be a sum of n.e.d. random variables,

$$T(g) = T_1(g) + T_2(g) + \ldots T_M(g)$$

and, the dwell time for a g-type platform can be a sum of n.e.d. random variables

$$T_D(g) = T_{D1}(g) + T_{D2}(g) + \ldots T_{DL}(g).$$

Such generalizations however can substantially increase the number of states needed to characterize the system.

STATE TRANSITIONS

Having identified and characterized the driving processes, it remains to characterize all of the state transitions. For each state, the possible predecessor states must be identified. That is, those states which could have <u>immediately</u> given rise to the current state under each of the (multi-dimensional) driving processes, must be found. In addition, the state transition probability flows, must be found. This is tractable using the Markovian assumptions for the driving processes. What we do is formulate the rules that govern the state transitions and determine the formulas for the corresponding transition flows. We get the computer to do the tedious work for all states.

Flows determined by parameters of the driving processes and the state transition RULES cause the cell to change state. Figure 4 shows a general state of a single cell. Not all states have all the flows shown. A cell is characterized by an <u>interconnection</u> of such states with the appropriate flows. In general, all flows are multi-dimensional. This is because new call arrivals and call completion flows can depend on platform type and on number of calls in progress on the platform, while hand-off arrival and departure flows depend on platform type and order of hand-off.

It is sometimes convenient to form a table that characterizes state transitions. An example is shown in Table 3 for case 5 (described previously), with C=8, and C_h=1. The ordered list of permissible states are shown on the left along with the number of channels in use and the number of communicating platforms. Columns further to the right are identified by a relevant driving process, under which is listed the immediate predecessor states that could have brought the system to the current state. Note, that in this example, since C=8 and C_h=1, no new calls are served in a cell if $j(s)>7$. Thus, the only flows into states 31-40, (for which $j(s)>7$) arise from a hand-off arrival driving process, {h}.

TABLE 3. STATE TRANSITIONS
G=1, N(1)=3, C=8, C_h=1, CHANNEL LIMIT CONSTRAINT ONLY

CURRENT STATE						PREDECESSOR STATES			
s	j	w	v_{11}	v_{12}	v_{13}	{n}	{c}	{h}	{d}
0	0	0	0	0	0	–	1	–	1,2,4
1	1	1	1	0	0	0	2,3	0	3,5,8
2	2	1	0	1	0	1	4,5	0	5,7,11
3	2	2	2	0	0	1	5,6	1	6,9,13
4	3	1	0	0	1	2	8	0	8,11,16
5	3	2	1	1	0	2,3	7,8,9	1,2	9,12,18
6	3	3	3	0	0	3	9,10	3	10,14,20
7	4	2	0	2	0	5	11,12	2	12,17,23
8	4	2	1	0	1	4,5	11,13	1,4	13,18,24
9	4	3	2	1	0	5,6	12,13,14	3,5	14,19,26
10	4	4	4	0	0	6	14,15	6	15,21,28
11	5	2	0	1	1	7,8	16,18	2,4	18,23,31
12	5	3	1	2	0	7,9	17,18,19	5,7	19,25,33
13	5	3	2	0	1	8,9	18,20	3,8	20,26,34
14	5	4	3	1	0	9,10	19,20,21	6,9	21,27,36
15	5	5	5	0	0	10	21,22	10	22,29,38
16	6	2	0	0	2	11	24	4	24,31
17	6	3	0	3	0	12	23,25	7	25,32
18	6	3	1	1	1	11,12,13	23,24,26	5,8,11	26,33
19	6	4	2	2	0	12,14	25,26,27	9,12	27,35
20	6	4	3	0	1	13,14	26,28	6,13	28,36
21	6	5	4	1	0	14,15	27,28,29	10,14	29,37
22	6	6	6	0	0	15	29,30	15	30,39
23	7	3	0	2	1	17,18	31,33	7,11	33
24	7	3	1	0	2	16,18	31,34	8,16	34
25	7	4	1	3	0	17,19	32,33,35	12,17	35
26	7	4	2	1	1	18,19,20	33,34,36	9,13,18	36
27	7	5	3	2	0	19,21	35,36,37	14,19	37
28	7	5	4	0	1	20,21	36,38	10,20	38
29	7	6	5	1	0	21,22	37,38,39	15,21	39
30	7	7	7	0	0	22	39,40	22	40
31	8	3	0	1	2	–	–	11,16	–
32	8	4	0	4	0	–	–	17	–
33	8	4	1	2	1	–	–	12,18,23	–
34	8	4	2	0	2	–	–	13,24	–
35	8	5	2	3	0	–	–	19,25	–
36	8	5	3	1	1	–	–	14,20,26	–
37	8	6	4	2	0	–	–	21,27	–
38	8	6	5	0	1	–	–	15,28	–
39	8	7	6	1	0	–	–	22,29	–
40	8	8	8	0	0	–	–	30	–

FLOW BALANCE EQUATIONS

To find the statistical equilibrium state probabilities for a cell, we write the flow balance equations for the states. These are a set of $s_{max}+1$ simultaneous linear equations for the unknown state probabilities, $p(s)$. They are of the form

$$\sum_{j=0}^{s_{max}} q(i,j) \, p(j) \;=\; 0 \;\;,\;\; i = 0,1,2,\ldots s_{max}-1 \qquad (1)$$

$$\sum_{j=0}^{s_{max}} p(j) \;\;=\; 1 \qquad \qquad \qquad (2)$$

in which for $i \neq j$, $q(i,j)$ represents the net probability flow into state i from state j, and $q(i,i)$ is the total flow out of state i. Flow <ins>into</ins> a state is taken as a positive quantity. Equation (2) requires that the state probabilities sum to unity. The total (state-to-state) flows, $q(i,j)$ are determined by the parameters of the underlying driving processes and can be found by considering the <ins>component</ins> of flow due to each.

ALGORITHM TO SOLVE EQUATIONS

The state probabilities can be determined using the following algorithm.

1. Given parameters, and assuming that {h} and {d} are uncoupled use a linear equation solver to find "tentative" state probabilities $p(s)$.
2. Find the average hand-off departure rates of various orders for all platform types.
3. Update the initial guess of average hand-off arrival rates of various orders for all platform types.
4. Iterate until there are no differences in average hand-off arrival and departure rates to within 4 significant figures.

5.   The final state probabilities are the equilibrium state probabilities that were sought.

DETERMINATION OF PERFORMANCE CHARACTERISTICS

Once the state probabilities are determined, the blocking and hand-off failure probabilities can be found using the following steps. First, identify those states which can give rise to blocking and the event(s) which cause blocking for each of those states. Then find the blocking probability using the values of p(s) and the probabilities of the events. Completely similar steps are used to determine hand-off failure probabilities.

CONCLUSIONS

Example performance characteristics were generated using the methodology described. These are shown in Figure 5. The parameters chosen are shown in Table 4. Consider the example performance characteristics for moderate traffic, say, Λ_0 = 6.0E-05 calls/sec. A change in C_h from 0 to 3 (increasing priority for hand-off calls) causes P_B to increase from 0.0026 to 0.065 (a factor of 25) while P_H decreases from 0.0026 to 1.6E-05 (a factor of 164). This indicates that cut-off priority can be effective in reducing forced terminations in a multiple call hand-off environment.

More importantly, the approach described here establishes a rich methodology which can be used to develop a broad range of realistic analytical models to characterize the hand-off problem in many scenarios of practical interest.

TABLE 4.   PARAMETERS FOR PERFORMANCE CHARACTERISTICS

G   = 1   one platform type

$N(1)$ = 3   ports per platform

C   = 8   channels per cell

v_{10} = 133   non-communicating platforms per cell

C_h = 0,1,2,3 channels per cell reserved for hand-offs

$\Lambda_n(1,0)$ = 3.0E-05 to 9.0E-05 calls/sec.
New call origination rate per port for non-communicating platforms.

$\overline{T}(1)$ = 100 sec.
mean unencumbered call holding time

$\overline{T}_D(1)$ = 400 sec.
mean dwell time of a mobile platform in a cell

α_{gi} = 2.0 ; g=1, i=1,2,3.
new call origination rate factor for a g-type platform with i calls in progress.
= $\Lambda_n(g,i)$ / $\Lambda_n(g,0)$

Constraints: $W(1) = C$, $J(1) = C$, $V_{max} = C$
CHANNEL LIMIT CONSTRAINT ONLY.

Note:
$\Lambda_n(1,0) = 6.0E-05$ calls/sec. corresponds to 22% of all ports on board non-communicating platforms in the cell <u>each</u> originating a call during one hour.

REFERENCES

[1] D.C. Cox and D.O. Reudink, "Layout and control of high-capacity systems," Chap. 7 in *Microwave Mobile Communications*, ed. by W.C. Jakes, Jr., John Wiley & Sons:New York, 1974, pp. 545-622.

[2] D. Hong and S.S. Rappaport, "Traffic model and performance analysis for cellular mobile radiotelephone systems with prioritized and non- prioritized hand-off procedures," IEEE Trans. Vehic. Technol., vol. VT- 35, pp. 77-92, Aug. 1986.

[3] S.A. El-Dolil, W-C. Wong, and R. Steele, "Teletraffic performance of highway microcells with overlay macrocell," IEEE J. Select. Areas Commun., vol.7, no.1, pp. 71-78, Jan. 1989.

[4] D. Hong and S.S. Rappaport, "Priority oriented channel access for cellular systems serving vehicular and portable radio telephones," IEE (British) Proceedings, Part I, Communications, Speech and Vision, vol. 136, pt.I, no.5, pp. 339-346, Oct. 1989.

[5] R. Guerin, "Queueing-blocking system with two arrival streams and guard channels," IEEE Trans. Commun., vol. COM-36, pp. 153-163, Feb. 1988.

[6] S.S. Rappaport, "The Multiple-Call Hand-off Problem in High-Capacity Cellular Communications Systems," Proc. IEEE Vehicular Technology Conference, VTC '90, Orlando, May 6-9, 1990, pp. 287-293. See Technical Report # 567, College of Engineering and Applied Sciences, State Univ. of New York, Stony Brook, N.Y. 11794.

181

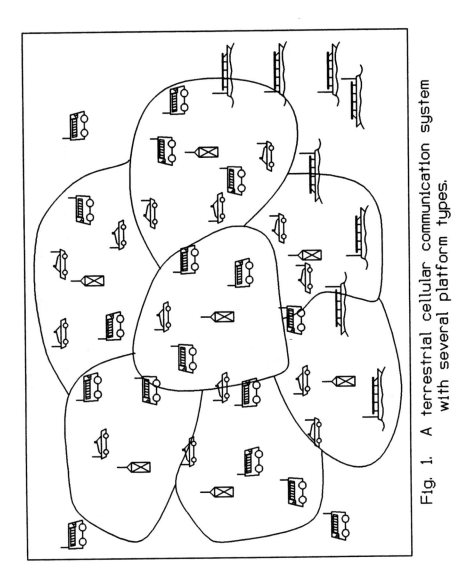

Fig. 1. A terrestrial cellular communication system with several platform types.

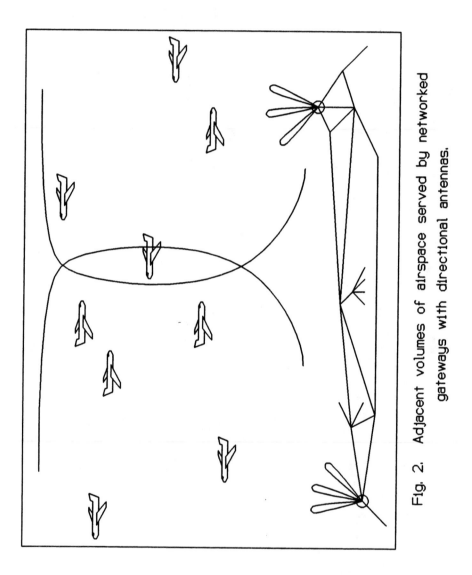

Fig. 2. Adjacent volumes of airspace served by networked gateways with directional antennas.

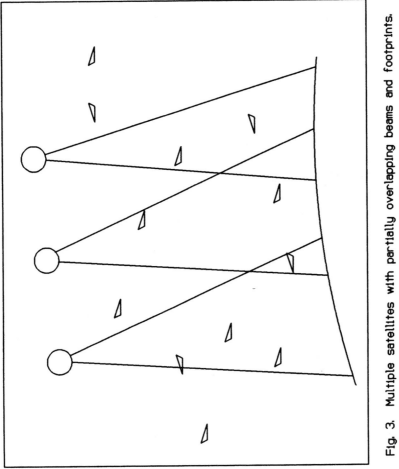

Fig. 3. Multiple satellites with partially overlapping beams and footprints.

184

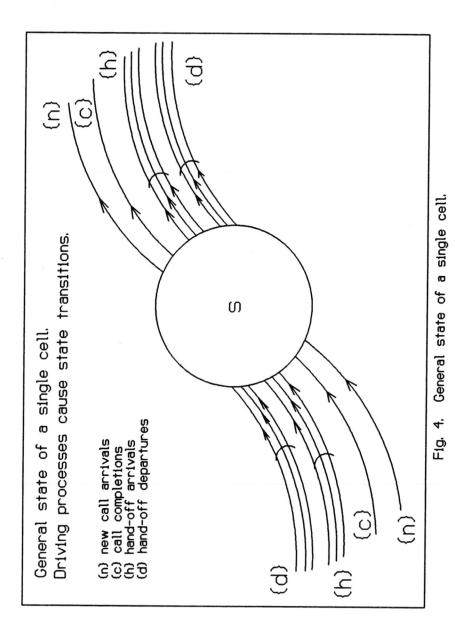

General state of a single cell.
Driving processes cause state transitions.

{n} new call arrivals
{c} call completions
{h} hand-off arrivals
{d} hand-off departures

Fig. 4. General state of a single cell.

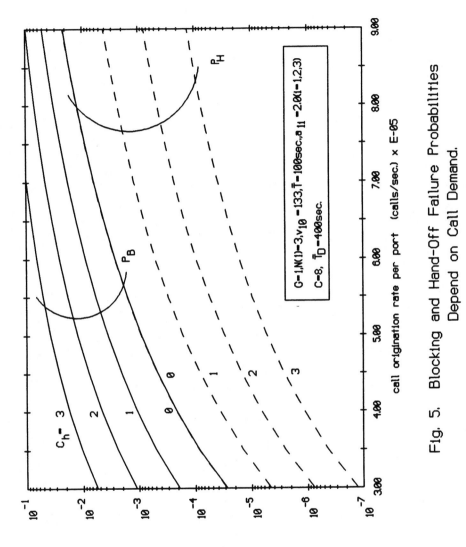

Fig. 5. Blocking and Hand-Off Failure Probabilities
Depend on Call Demand.

Handoff in Microcellular Based Personal Telephone Systems

Björn Gudmundson and Olle Grimlund
Ericsson Radio Systems AB
S-164 80 Stockholm, Sweden

Abstract

*When using small microcells with antennas mounted at lamp post level
the propagation characteristics will be very different from traditional large
cells. Firstly, the so called street corner effect, i.e. 20-30 dB drop of signal
level in 10-20 m, have to be considered. Very fast handoff algorithms and,
maybe, new strategies will be required. Furthermore the number of handoffs
per call will increase, due to the small cell size. Each handoff must therefore
be made with high success rate. Secondly, the propagation is very difficult
to predict since it depends to a large extent on the local environment. This
makes it very difficult to find a fixed reuse plan, i.e. dynamic channel
allocation (DCA) is probably required. In this paper we concentrate on the
handoff issue. Some discussion on the propagation characteristics and DCA
is however also made.*

Introduction

In future personal communication systems the demand for capacity will increase dramatically. This cannot be solved only by considering more effective
modulation methods, channel coding and speech coding. The solution to get
the wanted capacity is to introduce microcells, i.e. very small cells, so that the
channels can be reused more often.

The definition of a microcell used here is that the base station (BS) antenna is located at street level, for instance mounted 5-10 m up on a lamp post. Traditional BS antennas on the other hand are mounted very high (roof tops). Furtermore the emitted power will be on the order of milliwatts and the size of the microcells will be around 50-500 m.

One of the foreseen problems with microcells is the street corner effect, see [1,2]. As long as the mobile/portable station (MS) is in line-of-sight (LOS) the signal strength is high. When the MS is turning round a corner, i.e. Non-LOS (NLOS), the signal level drops very fast, 20-30 dB in 10-20 m. The C/I (carrier to interference ratio) might therefore also drop very fast. This phenomenon has to be considered when handoff-strategies and algorithms are discussed, see [4,5]

Another difficulty is the problem to predict the propagation [2], as will be shown later. This makes it very difficult to make a fixed channel allocation (FCA). Therefore dynamic channel allocation (DCA) schemes are probably required. When we say DCA we mean that the channel allocation is based on the instantaneous interference level, assuming that all channels are availible everywhere, see [3]. This is very different from the classical approach of DCA, where channels are moved according to nonhomogenous traffic demand.

The outline of the paper is as follows: In Section 2 some examples of real measurements from central Sockholm are shown, where especially the street corner effect is pointed out. Also the propagation models used in the handoff simulations are briefly described. In Section 3 we review the handoff strategies that are currently used, i.e. analog systems like AMPS, NMT and TACS, digital systems like GSM and US-TIA, and finally the digital cordless system DECT. We then try to describe DCA in more detail in Section 4. In Section 5 the handoff simulation model, which is based on a Manhattan street structure, is described. In Section 6 we show some results from simulations of Mobile Assisted Handoff (MAHO). A simplified GSM algorithm has been used as a basic MAHO-algorithm. Finally in Section 7 we try to make some conclusions and also describe the continued work that has to be done.

Propagation Measurements and Models

Extensive measurements both on 900 and 1700 MHz have been made in central Stockholm to analyze the propagation characteristics for microcells [2]. The examples described below are for 900 MHz. No main differences between the two carrier frequencies have been noted however. Except that the "fast" fading is of cource twice as fast for 1700 MHz compared to 900 MHz. In the examples the signal levels shown have been averaged over 6 m.

In Figure 1 the received signal strength for two different LOS streets are shown, i.e. the MS is in line-of-sight to the BS all the time. Note that the distance is on a logarithmic scale. Two conclusions can be drawn, firstly a break-point is noted (300-500 m) where the distance dependence changes from

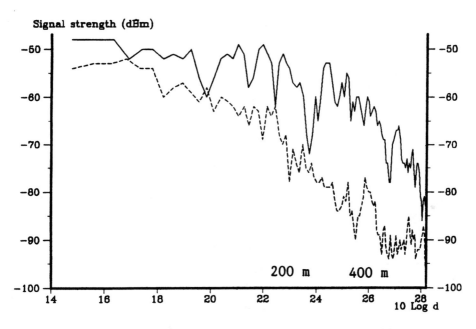

Figure 1: Received signal strength vs distance, two different LOS streets.

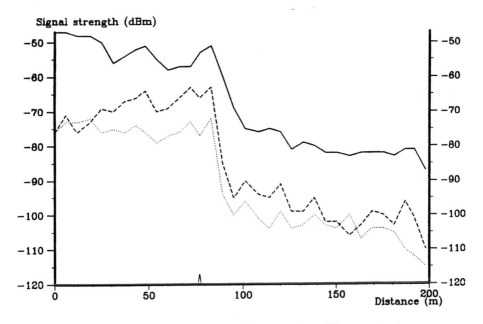

Figure 2: Received signal strength vs distance, three different street corners.

roughly d^{-2} to $d^{-2} - d^{-8}$. Secondly the path loss is very different from street to street, i.e. it is very difficult to predict.

In Figure 2 the street corner effect is shown. The three different curves are for three different corners, turning from the same street. A signal level drop of around 25 dB in 10 m can be noted. This is a typical street corner behaviour, very large discrepancies in other corners have been seen however. Again this makes a traditional fixed reuse plan impractical.

In the analysis we have tried to fit parameterized models to the measurement data, both for the path loss and the "slow" fading. Note that in all tested models the street corner effect is separately modeled as path loss, i.e. not as a part of the slow fading. This effectively reduces the standard deviation of the fading in the models to 3-5 dB.

The empirical model for the path loss that is used in the simulations is described in the Appendix. Basically on LOS streets it consists of two distance dependence factors, one before the break point and one after. Then the street corners are modeled with an imaginary transmitter located at the street corener, with the same distance dependence as for LOS. An example is shown in Figure 3 in Section 5.

It is not possible to describe the fading process in a traditional way, i.e. one slow component (log-normal) and one fast (Rayleigh). Instead all fading has a distribution similar to Rayleigh, but the dips are at a much larger distance than $\lambda/2$. The distinction between fast and slow fading is more diffuse than for large cells.

We have tried to analyze the correlation properties of the filtered (6m) fading process. This was done by applying AR-analysis. In fact an AR(1) process seems good enough, i.e. white noise filtered through a single pole low-pass filter. The correlation is larger further away from the BS (> 200 m). Roughly the correlation with a distance separation of 6 m is around 0.1 close to the BS (< 200 m) and around 0.5 far away (> 200 m).

Review of Handoff Strategies

When a MS is moving around during a call the signal level (or the C/I, if interference limited) might become unacceptably low **and/or** another BS can offer better quality. In that case one wants to transfer the call to a new channel (intracell handoff) or to a new BS (intercell handoff). The handoff strategies used in the analog cellular systems of today (AMPS, TACS, NMT), the digital cellular systems to be introduced 1991 (GSM in Europe, "TIA" in US) and the cordless systems to be introduced 1991 in Europe (DECT) are all quite different.

In AMPS (TACS, NMT) the MS is completely passive. The BS is supervising the quality of the current connection, i.e. measurements of RSSI and SAT-tone. Measurements (RSSI) of alternative connections are done by surronding BSs, by order from the MSC (mobile switching center). The decisions (when and where) are made at the MSC. The whole process can therefore be described as **network controlled handoff** (NCHO). Only intercell handoffs are possible. The handoff time (handoff needed to execution) is on the order of many seconds. One main drawback of NCHO is that the measurements of neighbouring channels can not be made very often. Therefore the accuracy is reduced. Furthermore to reduce the signalling load in the network the neighbouring BSs can not send measurement reports continously. This means that comparisons can not be made before the actual RSSI is below a certain threshold.

In GSM the handoff process is more decentralized. Both the MS and the BS are supervising the channel quality, i.e. RSSI and BER (bit error ratio). Measurements of alterative BSs is done in the MS (RSSI). The MS transmits the measurement results up to the BS twice a second. The decisions however are still made in the fixed network (BS/MSC). This method can be described as **mobile assisted handoff** (MAHO). Both inter- and intracell handoffs are possible. For GSM the handoff time is approximately 1s.

In the DECT system (Digital European Cordless Telephone) the handoff strategy is even more decentralized. The supervision of the own channel is done by both the MS (the portable) and the BS, i.e. RSSI and BER. In this case the BS transmits the measurements down to the MS. Measurement of alternatives are done by the MS, both BSs (RSSI) and free channels ("C/I"). In DECT also the decisions are made in the MS, and the method can therefore be described as **mobile controlled handoff** (MCHO). Both inter- and intracell handoffs are possible, in fact the intracell handoffs are the basis for DCA [3]. The handoff time for DECT is about 100 ms.

The trend is to decentralize the measurements and decision making, in order to get a fast handoff. Note that if all signalling on the air interface was error free there would not be any main differences in performance between MAHO and MCHO. The critical difference is that for MAHO a handoff command is transmitted from BS to MS. If that message is not received correctly the call is lost. Also if new BSs are not identified or recent measurement reports are missing, the handoff command might be delayed and therefore missed later. It is therefore interesting to study MAHO to see if it is feasible to use in the microcellular environment.

Dynamic Channel Allocation

Due to the problem of predicting the propagation in the microcellular environment, a fixed reuse pattern seems not feasible. Instead one wants a "self-planning" system, i.e. a system based on dynamic channel allocation (DCA). Note that the traditional DCA proposals, i.e. systems that dynamically can move channels between the cells based on traffic demand, needs the same planning as for FCA. Instead we want DCA schemes that base the channel allocation on the instantaneous interference situation. Note that all systems of course need careful planning of where to place the BSs.

Another important advantage of using DCA (based on the instantaneous interference situation) is the capacity gain that can be achieved, compared to an optimal fixed channel reuse plan. Gains of at least a factor 1.5 has been reported in the literature.

DCA based on the intereference situation will adapt not only to best C/I, it also automatically adapts to the traffic situation. One example of this type of system is DECT, see [6].

One claimed disadvantage of DCA is the probability of instability in the algorithm in overload situations. Our simulations on DECT have shown that this is not a problem [7]. The blocking increases of course, but the number of handoffs is still limited. Note that this is true for the DECT system, other proposals have to be tested in a similar way.

Simulation Environment

The handoff simulations are performed in a Manhattan street model, i.e. a recti-linear grid of streets and buildings. All streets and buildings are identical, and no elevation differences exists.

The propagation model, see Appendix, briefly described above is used. The "slow" fading is modeled as a log-normal process with standard deviation $\sigma = 3$ dB. The distance correlation is set to zero. The reason is that from the analysis it was found to be very low, at least close to the BS. If the MS speed is not too low, the assumption seems reasonable. Furthermore the fading from different BSs are uncorrelated.

In Figure 3 the received signal strength for two different routes in the Manhattan street structure are shown, for the described model. Note that the fading process is added to the signal in the simulations. For route B a 3 dB increase can be seen when passing the two middle crossings. The reason is that two similar NLOS paths exist and are added non coherently, i.e. an increase of 3 dB.

In the simulations so far we have only studied the case where the interference is modeled as a constant level. This is of course not the case for a real C/I limited system.

To test the handoff performance 2000 MS, with different fading, have driven the same route. Then different statistics have been collected, for instance when the handoff is made, the number of handoffs, handoff failure rate, etc.

In the simulations so far we have concentrated on MAHO, i.e. the decision is made in the fixed network, and the HO-command is signalled to the MS on the old channel. The algorithm is a stripped GSM-algorithm and therefore the performance results should not be interpreted as GSM-performance.

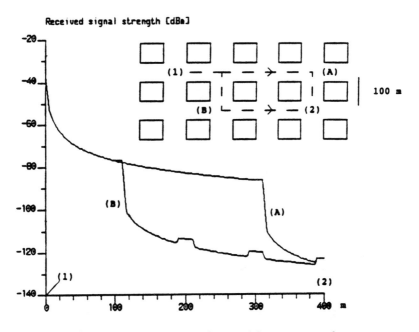

Figure 3: Example of propagation model, two routes shown.

The only detail in the protocol that can fail is that the handoff-command is not received correctly. Note however that the handoff-command is transmitted **34** times. It is assumed that all measurement results from the MS are received correctly. This is very important to note when looking at the results. Furthermore the BSIC (Base Station Identity Code) of the neighbouring BSs are assumed to be known in the MS. In a real GSM system this might also be a limiting factor.

The probability of receiving a handoff command correctly are derived from transmission simulations on the GSM system.

The only criterion used is a signal strength (RSSI) comparison, with a hysteresis level of 3 or 10 dB. The averaging length is set to 2 or 10 samples, i.e. 1 or 5 sec. The MS speed has been varied from 1 m/s up to 14 m/s (50 km/h).

We have only had two existing BSs in the simulations. In a real system we therefore also have the possibility to make handoff to a wrong BS.

Some Results for MAHO

Before we study the handoff performance some interesting C/I calculations are examined. Although FCA is not feasible for the microcellular environment, it

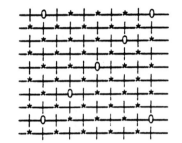

Figure 4: Fixed cell plan, clustersize = 8.

is interesting to study possible clustersizes etc. to get a feeling for the C/I problems. Here we just show an example for the cell plan in Figure 4. (0 means BSs with the same channel set.) The grid of BSs is quite dence, one BS in every other crossing. From the propagation data we understand that the most dangerous co-channel intereferers are located on LOS streets. On other streets however the channels can be reused more often, as indicated in the Figure.

In Figure 5 we have plotted the C/I distribution in the BS and the MS respectively, assuming the MS is connected to the same BS all the time. The MS started at the BS (0 m), moved to the first street crossing (100 m), the turned right/left to the next crossing (200 m). The different curves give the percentage of MS below a certain C/I value, at the corresponding distance

The large difference in behaviour in BS and MS is striking. The reason to the larger degradation in the BS is that the carrier (C) is affected by the street corner effect, while the interferer (I) is not (can still be in LOS). In the MS both the C and the I (most important) are subject to the same signal level drop, and therefore the C/I remains more or less constant after the crossing.

Now let's turn to the handoff performance, starting with the LOS case. In Figure 6 the MS route is shown, beginning at one BS and ending at another, on the same street. Note that a constant interference level is used in all handoff simulations. In this case the interference level was such that the C/I at the worst case point was 12 dB (mean), i.e. in the street crossing half-way between the two BSs. In all simulations there were really no problem with this case, except that sometimes many handoffs back and forth were made.

In Figure 7 we look at a case where the speed is very slow, only 1 m/s, and a very fast algorithm is used (averaging length $N = 2$ and hysteresis $\Delta = 3$ dB). In the Figure the number of handoffs/call are shown, where a call is equal to the time to drive the route in Figure 6. With this fast algorithm the handoff behaviour is quite unstable, most of the calls make more than 1 handoff. When the speed is higher and/or a slower algorithm is used, the problem becomes negligable.

Instead we turn to the more interesting case shown in Figure 8. This time the route includes a turn round a corner. The signal strength is shown in

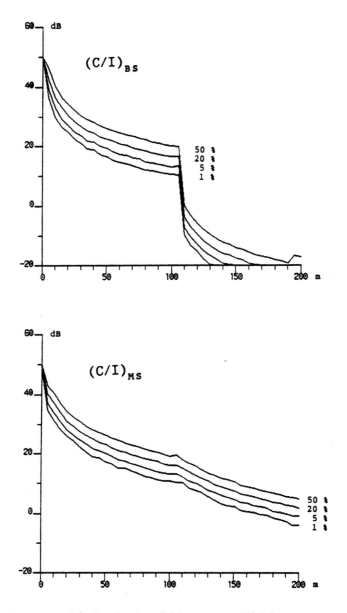

Figure 5: C/I distribution, LOS: 0-100 m, NLOS: 100-200 m.

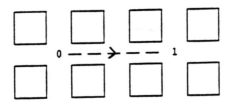

Figure 6: Route for LOS handoff.

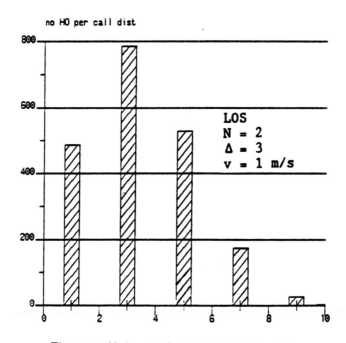

Figure 7: Nr handoff/call, LOS, fast algorithm.

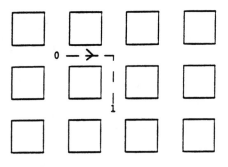

Figure 8: Route for NLOS handoff.

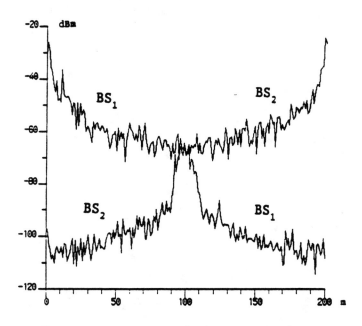

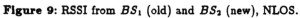

Figure 9: RSSI from BS_1 (old) and BS_2 (new), NLOS.

Figure 9, from the two BSs. This time we have put the intereference level so that the C/I in the crossing is 22 dB. Now, one can see that there is very short time for making the handoff decision.

In Figure 10 the number of handoffs/call is shown again, for two different speeds. First a very slow moving MS is shown. In this case no handoffs fail, instead we see the same behaviour as before, i.e. flipping from one BS to the other a couple of times. Note that if the algorithm is made slower (not shown) all calls are only making one handoff.

In the other case in Figure 10 we have increased the speed to 14 m/s (50 km/h), keeping the same algorithm. Now one can see that about 60 % of the calls are lost (number of handoff = 0).

In Figure 11 we have collected a number of different simulations. The two figures have different algorithms, the first is very fast ($N = 2$ and $\Delta = 3$), while the second is slow ($N = 10$ and $\Delta = 10$). What is shown is the probability for handoff success, with different MS speeds, versus C/I in the crossing. One can for instance see that increasing the speed from 7 to 14 m/s means that the C/I has to be increased about 4 dB to keep the same handoff performance (for the fast algorithm).

Again it should be noted that the simulations are made under very idealized situations. In fact only relative comparisons should be made so far. For instance the way we have modeled the interference situation is not realistic (constant level) for a C/I limited environment, see for instance Figure 5. Studies on "real" cell-plans have indicated a much better behaviour than shown above. The reason again is that the down-link is much better than the up-link, when "real" C/I situations are studied. Remember however that in a real handoff protocol the measurements are sent in the up-link.

Conclusion and Further Studies

The results presented here should be seen as an indication of what can be expected in the microcellular environment. The simulation details should of course be more refined, see below, before any firm conclusions can be drawn.

Some general statements can be made however. From the C/I calculations it is clear that the situation in the BS and in the MS can be very different. The non-reciprocity of C/I is more evident for microcells than for traditional large cells.

Handoff in the LOS case will naturally cause only minor problem. Some small concern has to be made however. Probably one want a fast handoff algorithm, for the NLOS handoff. Then the risk of getting many handoffs back and forth in the LOS case is high.

For the NLOS case, i.e when the MS is subject to the street corner effect, one can forsee problems. The handoff performance depends to a large extent on the parameters in the algorithm and on the MS speed. It is too early to

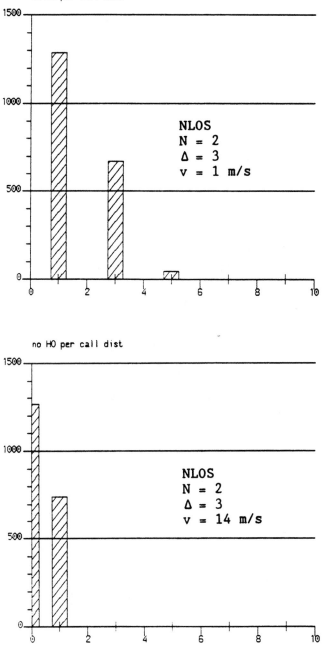

Figure 10: Nr handoff/call, NLOS, slow and fast MS-speed.

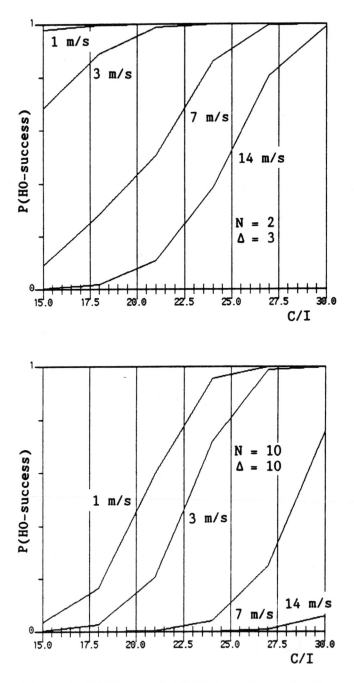

Figure 11: P(HO-success), NLOS, fast and slow algorithm.

state if MAHO is feasible or not, more detailed studies must be made. Perhaps the use of umbrella cells, see below, must be used for fast moving vehicles. Of cource if the clustersize is large one will always get good performance.

The work reported here is continued in different directions. At first more detailed studies of the real GSM handoff protocol will be made. We want to include more error sources in the protocol, for example: BSIC identification and errors in measurement reports.

It is also important to study more realistic environments. So far we have looked at constant intereference levels, that should be changed to real C/I scenarios, according to different cell-plans. Of course should also DCA be studied in the microcellular environment. Furthermore we should have more neighbouring BSs, to allow the possibility of selecting "wrong" BS.

One of the driving forces behind these studies was to investigate different handoff strategies [4], for instance MCHO. In the simulation scenario described in this paper, where only the handoff command can fail, MCHO will always succeed. Other means for characterizing the performance of MCHO must therefore be examined.

The use of umbrella cells to aid the handoff between microcells should also be studied. Perhaps this is the only solution for fast moving vehicles.

In this paper we have had no intention to optimize the parameters in the algorithm. For instance, how long should really the averaging time be? Furthermore there might be other more "intelligent" algorithms then just averaging.

Appendix: Empirical Propagation Models

The empirical model described here is based on measurements made in central Stockholm [2]. Basically it consists of two parts, one for LOS and one for NLOS propagation. In both cases a separation of the signal into three propagation factors is made, i.e. a global attenuation component dependent on the distance between transmitter and receiver, a log-normally distributed component of shadow fading and a fast fading component. The fast fading component is however assumed to be averaged out in the receiver.

In the simulation program only samples of the received signal strength is calculated, with a distance between samples equal to 1 m. Linear interpolation is then used to obtain the continuous signal.

To obtain the received power the following equation is used:

$$P_r = P_t - 32.4 - 20\log(f) - L(d) + F, \tag{1}$$

where P_r is the received power (dBm), P_t is the transmitted power (dBm), f is the used frequency (GHz), $L(d)$ is the attenuation (dB) as a function of distance d (m) and F is the log-normally disributed shadow fading (dB). Note that in the simulations $f = 1.7$ GHz. This equation gives the path loss between two isotropic antennas.

At first we start with the **global attenuation**, for the **LOS** case. $L(d)$ is calculated using the following equation:

$$L(d) = \begin{cases} k_1 \log(d) & d \leq D \\ (k_1 - k_2) \log(D) + k_2 \log(d) & d > D \end{cases} \qquad (2)$$

where d is the distance between the transmitter and the receiver and D is the distance to the breakpoint where the propagation characteristics changes, and k_1, k_2 describes the propagation law (with $k_1 = k_2 = 20$ we get free space propagation). Typical values for the parameters are: $D = 200 - 500$ m, $k_1 = 15 - 30$ and $k_2 = 40 - 80$. The values chosen in the simulations are: $D = 500$, $k_1 = 20$ and $k2 = 50$.

For the **global attenuation** of the **NLOS** propagation the idea is to have a fictitious transmitter located in the street crossing. This transmitter models the effect of reflections and diffractions around the corner. The output power from the fictitious transmitter is chosen to a value that makes the received power one meter from the fictitious antenna equal to the received power from the real antenna in the street crossing. A comparison with measurments, [2], shows that this model seems to fit quite well to the measured data.

The formulas (and the parameters) used for calculating $L(d)$ for NLOS propagation are the same as the formulas for LOS propagation. The attenuation is however zero from the center of the street crossing to the point where the transmitter actually gets out of sight, i.e. for a distance equal to half the street width. The distance d is here measured from the center of the street crossing to the receiver. To avoid an abrupt change in $L(d)$ when the transmitter, after passing the street corner, suddenly gets out of sight, the maximum derivative of L(d) is limited to a certain value (here 4 dB/m).

To continue the calculation of $L(d)$ for another NLOS street the procedure described above is simply repeated. The total attenuation from the transmitter to the receiver is then the sum (in dB) of the contributions from all the streets in between. If there are more than one possible path from the transmitter to the receiver where the signal can propagate, the total signal strength is given by an addition (in Watt) of the signal strength of each path.

Note the important fact that the largest contribution to the **shadow fading** is part of the global attenuation, i.e. the street corner. We will however model the extra randomness that is left as log-normal fading, F in the equation above. To account for the correlation between successive samples the fading is simulated using an autocorrelation function derived from measurements, [2].

To model the shadow fading one can put white gaussian noise through a one pole low-pass filter (pole location is a, see equation below), that is an AR(1) model. The autocorrelation is then given by the following equation:

$$R(k) = S a^{|k|}, \qquad (3)$$

where $R(k)$ is the autocorrelation between sample i and $i + k$, S is the fading variance (3 dB in all simulations), and a is the normalized correlation of the sampled fading process (also the pole location mentioned above). Then a is given by the following equation:

$$a = e^{r/d} \tag{4}$$

where e is the normalized correlation of the real fading process at distance d, and r is the resolution, i.e. the distance (in meters) between successive samples (in the simulation 1 m).

It turns out that the distance correlation is quite small, and will only have effect for very slow vehicle speed. In the simulations the correlation at 1 m has therefore been set to zero.

In a microcellular environment, the shadow fading of received signals from different transmitters located near each other, will probably be correlated and thus, a realistic model should take this into account. However, because of the lack of experimental data on this subject, that part has been omitted.

References

[1] S.T.S. Chia, R. Steele, E. Green and A. Baran, "Propagation and Bit Error Ratio Measurements for a Microcellular system", *J. IERE*, Vol. 57, No. 6 (supplement), pp 255–266, Nov./Dec. 1987.

[2] J-E. Berg et.al., *To be published*.

[3] H. Panzer and R. Beck, "Adaptive Resource Allocation in Metropolitan Area Cellular Mobile Radio Systems", *Proceedings VTC'90*, Orlando, Florida, May 6-9, 1989, pp 638–645.

[4] B. Gudmundson, "Handoff Strategies in Second and Third Generation Cellullar/Cordless Systems", *Proceedings WINLAB Workshop '89*, Piscataway, New Jersey, June 15-16, 1989.

[5] S.T.S. Chia and R.J. Warburton, "Handover Criteria for City Microcellular Radio Systems", *Proceedings VTC'90*, Orlando, Florida, May 6-9, 1989, pp 276–281.

[6] H. Ochsner, "Radio Aspects of DECT", *Proceedings DMR4*, Oslo, Norway, June 26-28, 1990.

[7] H. Eriksson and R. Bounds, "Performance of Dynamic Channel Allocation in the DECT System", Submitted for presentation at *VTC'91*, St. Louis, Missouri.

SIP Simulation for Urban Microcells

Jean-Frédéric Wagen
GTE Laboratories Inc.
40 Sylvan Road
Waltham, MA 02254

Abstract

The propagation simulation developed here is proposed to give insights, qualitative and quantitative information on radio coverage characteristics in urban microcell environments. The simulation is based on modeling the radio wave propagation using the Spectral Incremental Propagation (SIP) procedure. The geometry considered in this study consists of a rectangular layout of avenues intersecting with streets. The transmitter antenna height is well below the roof line of the surrounding buildings so that diffraction over the roofs can be neglected. The simulation results are compared with some published measurements. The expected guiding phenomenon along avenues and streets is demonstrated. For examples, similarly to measurements reported in the literature, the simulation results show peaks at intersections, on streets parallel to the street containing the transmitter.

Introduction

Microcells are becoming increasingly important as a strategy to increase capacity of cellular radio systems [e.g. Steele, 1985, Cox, 1989]. Because of their small dimensions, microcells reduce interference and may offer better communication characteristics than larger cells.

However, to achieve the expected capacity gain, the deployment of microcells must be carefully planned. System planners and radio engineers need tools, or at least guidelines, to help them design microcellular systems. Insights and information on radio coverage characteristics in urban microcell environments could be obtained through extensive measurements and field tests. The propagation simulation developed here is intended to complement or to be an alternative to those lengthy and expensive procedures.

The simulation is based on modeling the radio wave propagation using the Spectral Incremental Propagation (SIP) procedure [Cook et al., 1989], also known as the Beam Propagation Method in optics [Van Roey et al., 1981] and uses a technique similar to the Split- Step Method in Ocean Acoustics [Flatté, 1983] or the Multiple Phase Screen method in ionospheric propagation [Knepp, 1983; Kiang and Liu, 1985]. The SIP procedure uses Fast Fourier Transforms (FFT) to compute the wave propagation and diffraction. In the case considered here this procedure is an alternative to the more classical theories of diffraction (GTD, UAT or UTD [e.g., Felsen and Marcuvitz, 1973, Komyoumjian and Pathak, 1974, Chia et al., 1987]) and offers several advantages.

Uniform theories of diffraction are appealing since they provide an intuitive approach to diffraction phenomena and since a large amount of literature is available. However, their implementation can be computationally expensive especially if the wavefield must be known everywhere in the geometry considered. This is because the computations involve the evaluation of Fresnel integrals.

However, in the SIP procedure, efficient computation of numerous Fresnel integrals is accomplished essentially in parallel, using FFTs. Furthermore, the application of geometrical theories of diffraction requires the knowledge of which paths to keep and which paths to neglect. For example, in the modeling of microcell environments, Chia et. al [1987] choose to keep only the diffracted-diffracted, reflected-diffracted-diffracted, diffracted- reflected-diffracted and the reflected-diffracted- reflected-diffracted paths.

The simulation model presented here includes implicitly numerous diffractions and reflections.

The next section describes the SIP procedure. Then the application of this procedure to model radio wave propagation in a simplified microcell environment is described. The fourth section presents preliminary results of our simulation and comparisons with published measurements. Finally, a brief conclusion is presented.

The SIP Procedure

The Spectral Incremental Propagation (SIP) procedure [Cook et al., 1989] can be used to compute the forward scattered wavefield on a given plane some known distance away from an obstacle. The procedure iterates the following four steps to compute the wavefield through and after the obstacle:

1) compute the wavefield on the obstacle by multiplying the incident wavefield with a screen function describing the obstacle,

2) compute the spatial Fourier transform of this wavefield (angular spectrum [Ratcliffe, 1956]),

3) multiply this angular spectrum by a Fresnel propagator,

4) compute the inverse Fourier transform of this product to obtain the received field.

Letting the wavefield be denoted by $E(x, z)$, the equations corresponding to the above procedure are:

$$
\begin{align}
E_1(x, z_n) &= E(x, z_n) * T_n(x) \tag{1}\\
\hat{E}_1(\kappa, z_n) &= FT[E_1(x, z_n)] \tag{2}\\
\hat{E}_2(\kappa, z_{n+1}) &= \hat{E}_1(\kappa, z_n) * F(\kappa, z_{n+1} - z_n) \tag{3}\\
E(x, z_{n+1}) &= FT^{-1}[\hat{E}_2(\kappa, z_{n+1})] \tag{4}
\end{align}
$$

where

$$
F(\kappa, \Delta z) = e^{j\frac{\kappa^2}{2k_o}\Delta z} \tag{5}
$$

is the Fresnel propagator corresponding to propagation over a distance Δz. The distance Δz represents either the interval between the successive screens modelling the obstacle or the propagation distance after the obstacle. For completness, the received wavefield should be multiplied by the phasor $e^{-jk_o\Delta z}$.

For diffraction and reflection by building walls and assuming vertical polarization, $E(x, z)$ is conveniently taken as the electric field.

$FT[...]$ denotes the Fourier transform and is efficiently implemented using Fast Fourier Transform (FFT) techniques.

The screen function $T_n(x)$ represents the effects of the obstacle at z_n. When the obstacle is a conducting body, for example, a building, the screen function is zero over the width of the building and unity every where else, i.e., in the avenues adjacent to the building. Other examples of screen functions for conducting bodies are given by Cook et al. [1989]. Note that in wave propagation through a random media like the ionosphere, the screen function has no magnitude dependence but only a random phase variation [Knepp, 1983; Kiang and Liu, 1985].

Derivation of the equations (1)-(5) can be obtained from the Kirchoff-Huygens integral [Born and Wolf, 1964] by assuming a small incidence angle and small scattering angles. This is one reason why we choose to use the Fresnel propagator as given in (5) instead of the expression given by Cook et al. [1989] (i.e., $exp[-j\sqrt{(k_o^2 - \kappa^2)\Delta z}]$). Note that under the approximation of small scattering angle and adding the phasor mentionned above, the two expressions are equal. Another reason is that $exp[-j\sqrt{(k_o^2 - \kappa^2)\Delta z}])$ causes numerical problems for large scattering angles (i.e., large κ).

The implementation of the SIP procedure using FFT techniques requires the sampling of the wavefield in the x direction. The sample interval Δx must be smaller than a wavelength. The sampling interval $\Delta \kappa$ in the wavenumber domain is $2\pi/L$ where $L = N\Delta x$ is the total width of the screen function (and N is the number of points in the FFT). Sampling of the Fresnel propagator is adequate if the variation in the phase of the Fresnel propagator is less than π.

This requirement leads to the following inequality:

$$L > \frac{\lambda \Delta z}{\Delta x} \tag{6}$$

Another inequality is obtained by choosing the maximum wavenumber κ_{max} $(= N/2\Delta\kappa)$ to be larger than k_o times the angle subtended by the first Fresnel zone F_1, i.e.,

$$(N/2)\Delta\kappa > k_o(F_1/\Delta z). \tag{7}$$

When $F_1 = \sqrt{\lambda \Delta z}$, i.e., for $\Delta z \gg \lambda$, then (7) leads to

$$\Delta x < \frac{\sqrt{\lambda \Delta z}}{2}. \tag{8}$$

This condition is already satisfied by choosing Δx smaller than the wavelength λ.

For the case of a half infinite screen (i.e., $T(x) = 0$ for $x < 0$ and $T(x) = 1$ otherwise), it is found that the SIP procedure gives accurate results only for $x > -\lambda \Delta z/(2\Delta x)$, where Δz is now taken as the propagation distance after the screen. This is expected since the maximum angle correctly reproduced in the simulation is given by $(N/2)\Delta\kappa$ which multiplied by Δz corresponds to a transverse distance x. Thus, the sampling interval Δx must be chosen appropriately small to obtain accurate results. In the simulation results presented in section 4, the sampling interval Δx is chosen to be approximately a third of the wavelength λ. The screen size is easily chosen to satisfy (6) and depends principally on the number of streets or avenues that need to be considered. The interval Δz between screens is chosen equal to Δx.

Modelling of Microcell environments

The geometry considered in this study consists of a rectangular layout of avenues intersecting with streets (Figure 1). The transmitter antenna height is well below the roof line of the surrounding buildings so that diffraction over the roofs can be neglected. Ground scattering and reflections are also neglected. These assumptions are made since coverage in the non-line-of-sight cases is the main interest in this investigation. Propagation models for the line-of-sight cases have been proposed recently by Harley [1989] using an empirical model, by Green [1990], using a crude 4-ray model, and by Kozono and Taguchi [1990], using the Fresnel-Kirchhoff diffraction equation which is similar to the Kirchoff-Huygens integral used implicitly in our model. However, in the simulation presented here, only simplified assumptions were made to evaluate the field in line-of-sight, i.e., in the street (or avenue) in which the transmitter is located.

To compensate for the inaccuracy caused by the small scattering angle assumptions, the computations are performed separately along the avenues and along the streets (Figure 1). Then, the resulting field strength is assumed to be the average of the two contributions resulting from the computations along the two perpendicular directions.

To introduce some details of the numerical simulation, let us consider a transmitter located in a street as represented in Figure 1 (top panel). For the propagation along the avenues, the initial field is computed assuming only free space from the transmitter to the beginning of each avenue. That is the initial field at $s = sw/2$ (where sw is the width of the

212

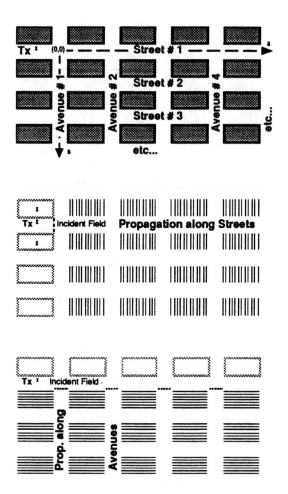

Figure 1: Geometry for the modeling of radio wave propagation in urban microcell environments. The top panel shows the notation used. The mid and bottom panels show graphically how the SIP procedure is applied (see section 3 for details).

streets) equals:

$$E(a, sw/2) = \frac{\sqrt{P}}{k_o r} \; e^{-jk_o r} \tag{9}$$

where

$$r = \sqrt{(s_{srce} - sw/2)^2 + (a_{srce} - a)^2} \tag{10}$$

and (a_{srce}, s_{srce}) give the coordinates of the transmitter location (Figure 1) and P is a constant representing the transmitted power. In this investigation, omnidirectional antennas are assumed, thus the effects of the antenna pattern are not included in (9).

For propagation along the streets (direction a), the initial field is computed in two steps. First, only free space propagation is assumed from the transmitter to the end of the street where the transmitter is located. Then, the field incident at the beginning of each street is computed using the SIP procedure (with Δz equal to the avenue width). This accounts for diffraction effects of the building wedges at the end of the street in which the transmitter is located (Figure 1, mid panel). To account for reflections due to the building walls on either side of the transmitter, the contributions of 2 mirror sources are added (see the two extra × marks in the mid panel of Figure 1).

Note that because of the inaccuracy of the SIP procedure for $|s|$ (or $|x|$) $< \lambda \Delta z / (2 \Delta x)$, the computation of the initial field for the propagation along streets uses a sampling interval Δx smaller than in the remaining of the computations.

The effects of the buildings are accounted for in the SIP procedure by letting the screen function be unity in the avenues or streets and zero in the buildings. The screen functions are represented graphically in Figure 1 by the series of parallel line segments. The screen function is zero on the solid line and unity elsewhere. Reflections by the building walls are implicitly computed when the SIP procedure is iterated over the very short interval Δz between screens (which is taken smaller than a wavelength).

To clarify the result of the computations, averages are made over three fourth of the transverse distance of the avenue or street to give only a single value of field strength versus the distance in a particular street or in a particular avenue.

In the following section, a few results from the simulation are presented and are compared to some published measurements.

Results and Comparisons with Published Measurements

Figures 2, 3 and 4 show the simulation results for a somewhat typical case. The radio wave frequency is 900 MHz. The width of the avenues and the width of streets are considered to be the same and equal to 35 m. The buildings are assumed to be 90 m by 120 m, with their largest width along the streets. The transmitter is located in the first street, 1.5 m from the left building wall (left looking towards the first avenue) and 102.5 m away from the intersection with the first avenue. This replicates approximately the microcell environment considered by Chia et al. [1987] and to a lesser extend, the urban area considered by Whitteker [1988].
Figure 2 presents the field strength in three adjacent streets versus the distance along each street. Note that the small fluctuations in the figures (as well as for example the narrow

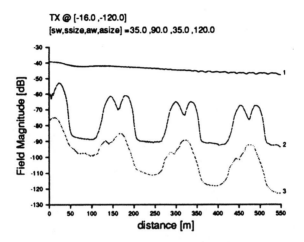

Figure 2: Magnitude of the wavefield in the first 3 streets (street number given as a parameter) versus distance along each street. The locations of the peaks in the field strength correspond to the intersections with avenues. The geometry is approximately similar to the urban microcell environment considered by Chia et al. [1987].

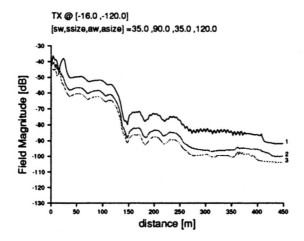

Figure 3: Magnitude of the wavefield in the first 3 avenues (avenue number given as a parameter) versus distance along each avenue. The locations of the peaks in the field strength correspond to the intersections with streets. The geometry is the one considered for Figure 2.

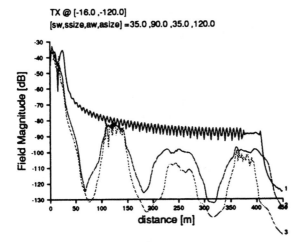

Figure 4: Magnitude of the wavefield in the first 3 avenues (avenue number given as a parameter). Only the propagation along the streets is considered (unlike Figure 3 where contributions from the propagation along the avenues is also taken into acount). The geometry is as for Figures 3.

dip in the larger peak) are due to the numerical analysis of the diffraction phenomena and should not be considered as measurable in real environments.

In Figure 2, 15 to 30 dB peaks at every intersection are readily observed. Figure 2 compares favorably with the measurements made by Chia et al. [see Figures 9 and 20 in Chia et al., 1987] or Whitteker [1988] although the decrease in signal strength from one avenue to the next is much smaller then the measured values. This is may be because the free space approximation for the initial field (in street number 1) is too optimistic in that it leads to too large values for the initial field. Another possible explanation is that in the present implementation of the SIP procedure, reflections from building walls are taken into account although assuming less than perfect reflections is probably more realistic.

In streets further away from the transmitters, (e.g. street # 3 in Figure 2), the maxima encountered at the intersection broaden, because more energy leaks from the avenues into the streets and also because more energy is coupled from the first avenue into the streets. It is then not surprising that for streets even further away from the transmitter and for some geometries, the energy coupled from the first avenue into the streets becomes very small and the field strength is instead mainly due to energy coupled from the first street into the avenues. Such conditions lead to sharp peaks at the intersections as can be seen in Figure 5, which will be explained later.

Figure 3 presents the field strength for the direction perpendicular to the one considered for Figure 2. That is, Figure 3 presents the field strength in three adjacent avenues versus the distance along each avenue. When the transmitter is deep inside a street, the field strength in the first avenue is mainly due to diffraction by the building wedges at the intersection (of street #1 and avenue #1). This explains the sharp peak from line-of-sight (LOS) to non LOS at the intersection of the first avenue and first street. The energy is then guided by the avenue and leaks into the streets. This behavior can explain the plateau shown in Figure 3. However, published measurements usually fail to show such plateau, showing much sharper decay in the first avenue. That behavior can be reproduced in our simulation if only the contribution from the propagation along streets is considered (as shown in Figure 4). This is an indication that further work is needed to improve on the free space assumption when evaluating the initial field for the propagation along avenues. Although crude, the free space assumption can be used to obtain an initial understanding of the scattering processes into the avenues and streets.

Note in Figure 4, that the field in the first avenue (#1) drops suddenly at about 400 m because of the numerical inaccuracy in the SIP procedure mentioned earlier. It is verified that this particular distance is given from the relation $\lambda \Delta z / (2 \Delta x)$ with the wavelength $\lambda = 0.33 m$, the propagation distance Δz taken as the width of the first avenue (35 m) and the sampling interval $\Delta x = 0.0148$.

Figure 5 shows similar results as Figure 2 but for another geometry. Namely, the widths of the avenues and streets are taken to be 30 m (instead of 35 m for the case of Figure 2). The building dimensions are 30 by 100 m (instead of 90 by 120 m). The transmitter is located in the first street, 5 m away from the building wall and 30 m away from the intersection with the first avenue. Those dimensions were chosen to replicate approximately the microcell environment considered by Kaji and Akeyama [1985]. Note that in their paper, Kaji and Akeyama present a 3-D graph of received

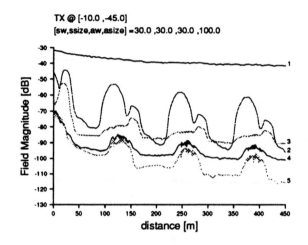

Figure 5: Magnitude of the wavefield in the first 6 streets (street number given as a parameter) versus distance along each street. The geometry is approximately similar to the urban microcell environment considered by Kaji and Akeyama [1985]. For comparison purpose, see Figure 3 of Kaji and Akeyama [1985].

power versus distance along avenues and streets which seems to show a similar behavior to the one observed in Figure 5 (with peaks at intersection becoming shallower, then sharper again), as explained earlier in this section.

The effects of transmitter location is shown in Figures 6 and 7. For Figure 6, the transmitter is located closer to the intersection with the first avenue (10 m away instead of 30 m). The increased coupling of energy into the first avenue and then into the streets causes the shallower peaks. Figure 7 shows the case where the transmitter is actually inside the intersection. In this case, the behavior of the field strength resembles the one shown in Figure 4. This is because the field strength is dominated by the free space field incident at the beginning of each street and avenue. Finally, Figure 8 (and

221

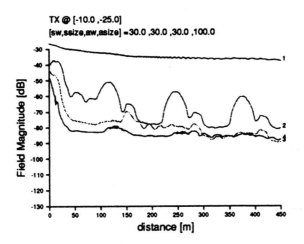

Figure 6: Case similar to the one presented in Figure 5 but for a transmitter located closer to the intersection of the first street with the first avenue (10 m away instead of 30 m away).

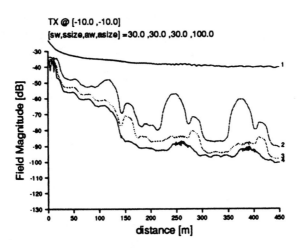

Figure 7: Case similar to the one presented in Figure 5 but for a transmitter located in the intersection of the first street with the first avenue.

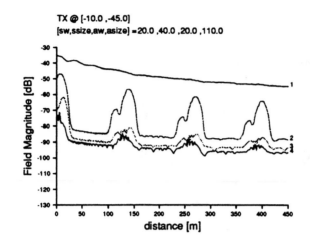

TX @ [-10.0 ,-45.0]
[sw,ssize,aw,asize] =20.0 ,40.0 ,20.0 ,110.0

Figure 8: Case similar to the one presented in Figure 5 but for narrower avenues and streets (20 m wide instead of 30 m wide).

Figure 9) show the cases of narrower (and wider) avenues and streets. Again the results can be explained intuitively from the point of view of coupling of the radio wave energy into the avenues and streets: narrower width thus less coupling thus sharper peaks (Figure 8); wider width thus more coupling thus wider peaks (Figure 9).

Conclusion

A theoretical model using the SIP procedure is proposed for the computation of wave propagation and diffraction in microcellular environments. The effects of the transmitter locations and the effects of avenue/street widths are briefly investigated. The expected guiding phenomenon along avenues and streets is demonstrated. The simulation results

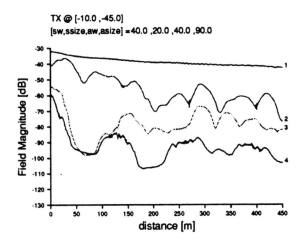

TX @ [-10.0 ,-45.0]
[sw,ssize,aw,asize] = 40.0 ,20.0 ,40.0 ,90.0

Figure 9: Case similar to the one presented in Figure 5 but for wider avenues and streets (40 m wide instead of 30 m wide).

show 15-30 dB peaks at intersections, on streets parallel to the street containing the transmitter when the transmitter is well inside that street. Similar values have been reported in the literature. As expected, these peaks decrease in size when the transmitter is located within an intersection instead of in a street.

The simulation results reveal variations along avenues and streets that are qualitatively similar to published measurements. A better model of the initial field is required to improve the simulation and better fit the measurements. Improvements are probably also possible if the implementation of the SIP procedure is modified to account for the effects of partial or no reflection by the building walls. Further work is also required to analyze more complex geometries, for example without the regular rectangular layout considered here and possibly with random variations. Finally further comparison with propagation measurements is needed to validate the model.

References

Chia, S. T. S., R. Steele, E. Green, and A. Baran, Propagation and bit error ratio measurements for a microcellular system, *Jour. IERE* Vol. 57, No. 6 (Suppl.), pp. S255-S266, Nov./Dec. 87.

Cook, G. G., A. P. Anderson, and A. S. Turnbull, Spectral incremental propagation (SIP) procedure for fast calculation of scattered fields from conducting bodies, *IEE Proc.* Vol. 136, Pt. H, No.1, p. 34-38, Feb 89.

Cox, D. C., Portable digital radio communications-An approach to tetherless access, *IEEE Com. Magazine*, pp. 30-40, July 1989.

Felsen, L. B., and N. Marcuvitz, Radiation and Scattering of Waves, *Prentice-Hall* Inc., Englewood Cliffs, New Jersey, 1973.

Flatté, S. M., Wave propagation through random media: Contributions from ocean acoustics, *Proc. IEEE* 71, pp. 1267-1294, 1983.

Harley, P., Short distance attenuation measurements at 900 MHz and 1.8 GHz using low antenna heights for microcells, *IEEE Jour. SAC-7 (1)*, pp. 5-11, Jan. 1989.

Kaji, M., and A. Akeyama, UHF- band radio propagation characteristics for land mobile system using low antenna height base stations, *IEEE AP-S 85*, pp. 835-838, 1985.

Kiang, Y.-W., and C. H. Liu, Multiple phase-screen simulation of HF wave propagation in the turbulent stratified ionosphere, *Radio Science* 20 (1), pp. 652-668, 1985.

Knepp, D. L., Multiple phase- screen calculation of the temporal behavior of stochastic waves, *Proc. IEEE* 71 (6), pp. 727-737, 1983.

Komyoumjian, R. G., and P. H. Pathak, A uniform geometrical theory of diffraction for an edge in a perfectly conducting surface, *Proc. IEEE-62 (11)*, pp. 1448-1461, Nov. 1974.

Kozono, S., and A. Taguchi, Propagation characteristics of low base-station antenna on urban roads, *Electronics and Communications in Japan*, part 1., Vol. 73, No. 1, pp. 75-85, 1990.

Ratcliffe, J. A., Some aspects of diffraction theory and their application to the ionosphere, *Rep. Prog. Phys.* 19, pp. 188-267, 1956.

Steele, R., Towards a high- capacity digital cellular mobile radio system, *Proc. IEE Pt. F - 132 (5)*, pp. 405-415, Aug. 1985.

Van Roey, J., J. van der Donk, and P. E. Lagasse, Beam propagation methods: Analysis and assessment, *J. Opt. Soc. America 71*, p. 803, 1981.

Whitteker, J. H., Measurements of path loss at 910 MHz for proposed microcell urban mobile systems, *IEEE Trans. VT-37 (3)*, pp. 125-129, Aug. 1988.

Evaluation of VTS CSMA for Media Access Control in Land Mobile Data Communication

David Chi-Yin Chan
Motorola Mobile Data Division
11411 No. 5 Road, Richmond,
B.C. Canada V7A 4Z3

Abstract

In land mobile data communication networks the Media Access Control facility regulates mobile data terminal access to the radio channel. In selecting the mobile radio channel access protocol, the emphasis should be placed on fair access for packet transmission among the mobile data terminals, with the objective of maximizing the throughput-delay performance. Given the severity of the noise characteristics for the mobile radio channel, the access protocol must provide a consistent high level of performance despite the inherent unreliability of data transmission. And, finally, due to the processing and memory limitations of a cost effective mobile data terminal, the access protocol must also be relatively simple to implement.

The Virtual Time Synchronous Carrier Sense Multiple Access (VTS CSMA) technique has been shown to provide fair access and optimum performance for shared channels [1] [4] [5]. VTS CSMA is appealing for mobile data applications because it supports fair access through a distributed FCFS queuing discipline that is not dependent upon centralized control information, and is a relatively simple algorithm to implement.

This paper examines the application of VTS CSMA to mobile radio channel access, and describes a proprietary, digital version of the technique referred to as Virtual Time Synchronous Digital Sense Multiple Access (VTS DSMA). The expected throughput-delay performance for a typical mobile data communications network is evaluated under varied traffic load conditions. Consideration is given to the selection of the re-transmission algorithm, and a scheme for prioritizing packet transmissions. It is shown that the overall performance of VTS DSMA is relatively insensitive to the effects of data transmission errors, and virtual time clock jitter and drift.

1. Introduction

In general, land mobile data communications involves the transfer of user information over a RF channel, between two or more moving data terminals. Commercial systems usually require access to fixed network services, and thus have a "star" shaped system topology, where each of the mobile terminals communicates with a fixed base station, as illustrated in Figure (1). The communications path between base and mobile is a full duplex link, with one radio channel, the outbound channel, supporting data transfers from base to mobile, and a second radio channel, the inbound channel, supporting data transfers from mobile to base. On outbound channel, the packet switch can employ a Time Division Multiplex Access (TDMA) protocol while the distributed terminals require a random access protocol on the inbound channel. Among the variety of random access protocols available, Virtual Time Synchronous Carrier Sense Multiple Access (VTS CSMA) provides for fair packet transmissions on a First-Come-First-Serve (FCFS) basis, the lowest probability of collisions, and the best throughput-delay performance[1] [4] [5].

The operation of VTS CSMA requires a "real time" clock and a "virtual time" clock at each terminal [4]. These clocks are discrete and advance in units of slots; the real time clock advances continuously, while the virtual time clock advances only when the inbound channel is idle. When advancing, the virtual time clock is incremented at two different rates relative to the real time clock: when virtual time lags behind real time, the virtual time clock is advanced at a rate η times the real time clock rate, $\eta > 1$. When virtual time has caught up to real time, the virtual time clock is advanced at the same rate as real time or, equivalently, at a rate $\eta = 1$. The factor η is referred to as the "virtual time clock rate".

If a packet to be transmitted on the inbound channel arrives when virtual time is caught up to real time, the packet is transmitted immediately at the next slot boundary. If virtual time lags real time when a packet arrives, the packet will be marked with the current real time value, the so called "stamped time", and held until virtual time catches up to this stamped time.

This paper examines the performance of VTS CSMA over a mobile radio channel and analyses the effects of retransmission control, channel noise, clock jitter and drift, and busy/idle channel detection errors.

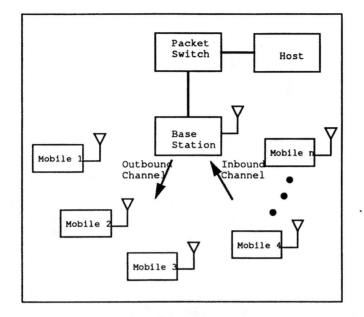

Figure 1: An example of Land Mobile Communications Networks

2. VTS CSMA

The VTS CSMA protocol always yields better throughput and delay performance than other CSMA protocols [1] [5]. The throughput, S, of the VTS CSMA protocol is given as follows:

$$S = E[H_s] / E[L_s] \qquad (1)$$

where $E[H_s]$ is the expected amount of useful work and $E[L_s]$ is the expected size of a random slot [4].

In a bandlimited and unreliable mobile radio channel, the critical measurement of performance is the achievable throughput as a function of the attempted transmissions. The throughput of the synchronous non-persistent CSMA protocol is given as [2]:

$$S = \frac{a * G * e^{-a*G}}{(1+a)(1 - e^{-a*G}) + a} \qquad (2)$$

where a is the normalized collision interval and G is the attempted traffic load (packets/packet duration).

The transmission rate, H, of synchronous non-persistent CSMA is given as:

$$H = \frac{a * G * (2 - e^{-a*G})}{(1+a)(1 - e^{-a*G}) + a} \qquad (3)$$

The throughput of the asynchronous non-persistent CSMA protocol is given as [2]:

$$S = \frac{G * e^{-a*G}}{G * (1 + 2 * a) + e^{-a*G}} \qquad (4)$$

The transmission rate, H, of asynchronous non-persistent CSMA is given as:

$$H = \frac{G * (1 + a * G)}{G * (1 + 2 * a) + e^{-a*G}} \qquad (5)$$

With VTS CSMA, an arrived packet is transmitted when virtual time passes real time. No transmission rescheduling is required when the channel is busy;

thus, the attempted traffic, G, for equation (1) is the same as the transmission rate, H. The throughput and transmission rate relationship is more relevant while comparing throughput performance among VTS, synchronous and asynchronous, non-persistent CSMAs.

Figure (2) compares the throughput performance between VTS CSMA and two other CSMA protocols. The packet duration, T, is 0.04 second, and the collision interval, τ, is 0.008 seconds for land mobile applications. A virtual time clock rate of 4 is selected as the optimal value for average packet durations ranging from 0.04 to 0.12 seconds. In this case, VTS CSMA yields a throughput capacity 5% and 21% higher than synchronous and asynchronous, non-persistent CSMA, respectively.

Simulations were performed to examine the delay performance of VTS CSMA and asynchronous, non-persistent CSMA. At an average retransmission delay of 0.75 second, and an average acknowledgment timeout of 0.02 second. Figure (3) compares the simulation results of delay performance with the range of normalized throughput levels from 0.05 to 0.40, and it shows VTS CSMA providing shorter average delays than asynchronous, non-persistent CSMA. For example, at a throughput level of 0.3, VTS CSMA achieves an average packet transfer delay shorter than CSMA.

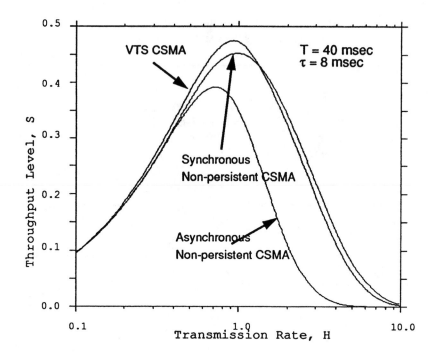

**Figure 2: Comparing Throughput and Transmission Rate Relationship
between VTS CSMA and Non-persistent CSMAs**

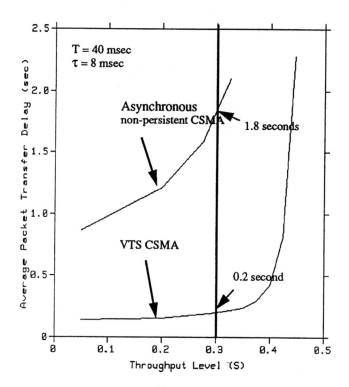

Figure 3: Delay vs Throughput Between VTS and Asynchronous Non-persistent CSMA

3. Re-Transmission Algorithm

In land mobile data communications, the packets making up the traffic load have varied durations. The adaptive geometric re-transmission algorithm provides a simple and effective method for maximizing throughput capacity and minimizing packet transmission delays by selecting the re-transmission probability according to packet duration. This performance of VTS CSMA with this re-transmission algorithm is relatively insensitive to the change of the average message duration.

The re-transmission probability, f, for the geometric re-transmission algorithm is defined as follows [1]:

$$f = \alpha / N_b \qquad (6)$$

where the re-transmission coefficient, α, is defined as:

$$\alpha = G - \lambda * \eta \qquad (7)$$

and the average number of blocked stations, N_b, is defined as:

$$N_b = W_r * \lambda \qquad (8)$$

W_r is the average waiting time due to re-transmissions.

With a fixed collision interval and throughput capacity, the optimal re-transmission probability depends on the packet duration. A static re-transmission algorithm uses a fixed re-transmission probability, optimized to the average packet duration, while the adaptive re-transmission algorithm assigns a different re-transmission probability for each packet retransmission according to the packet duration.

With a geometric traffic load and a throughput level at 80% of the capacity, simulations were performed to analyze the delay performance of VTS CSMA with the static and adaptive re-transmission algorithms. When the average packet duration is known, figure (4) shows similar delay performance of VTS CSMA with the adaptive re-transmission algorithm over that with the static re-transmission algorithm.

The average packet duration in a land mobile communications system varies frequently, the re-transmission probability of the static algorithm may be deviated from the optimal value. With 50% off from the optimal re-transmission probability of the static algorithm, figure (4) compares the delay performance of VTS CSMA with the adaptive re-transmission algorithm over that with the static re-transmission algorithm. The adaptive re-transmission algorithm reduces the average packet transmission delay by about 10% over the static re-transmission algorithm when the packet duration is less than 0.12 seconds. The adaptive re-transmission algorithm provides about the same average packet transmission delay as the static re-transmission algorithm when the packet duration is greater than or equal to 0.12 seconds.

From the above analysis, the adaptive re-transmission algorithm has been shown to have a small improvement on the delay performance over the static re-transmission algorithm when the re-transmission probability is off by 50% of the optimal value. Nonetheless, due to its simplicity, the adaptive algorithm is preferred over the static algorithm.

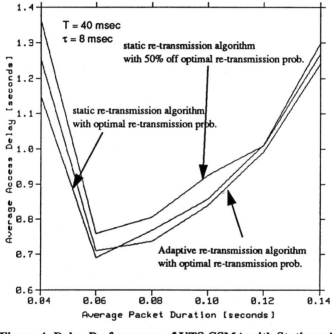

Figure 4: Delay Performance of VTS CSMA with Static and Adaptive Re-transmission Algorithms

4. Packet Transmission Errors

In a real mobile data communications channel, packet transmissions fail as a result of channel induced errors[6], and the throughput/delay performance degrades as a result. For example, at a five percent transmission failure rate due to channel errors, the access delay increases by 280%, for traffic loads near capacity, as illustrated in figure (5).

However, by allowing a small margin between the allowable traffic level and the available capacity, channel stability can be ensured, even for higher packet transmission failure rates.

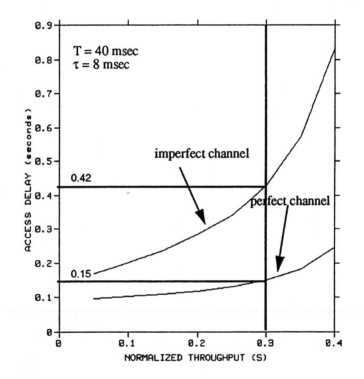

Figure 5: Average Packet Transmission Delay vs Throughput Level for Perfect and Imperfect Channels

5. Virtual Time Clock Jitter and Drift

The virtual time clock synchronization is important to the performance of VTS CSMA because local clock references vary from instant to instant due to sampling variations - jitter and long term thermal variation and component inaccuracies - drift.

Simulations were performed to analyze the delay performance of VTS CSMA with different virtual time clock jitter levels. Table (1) shows the delay performance of VTS CSMA with virtual time clock jitter levels of 0, 10^{-4} and 10^{-3} seconds, and corresponding packet transfer delay increase of 12% and 37%, respectively.

Jitter Noise [seconds]	Packet Transfer Delay [sec]	
0.0000	0.174	$T = 30$ ms $\tau = 8$ ms $\eta = 4$ $S = 0.30$
0.0001	0.195	
0.0010	0.238	

Table 1: Packet Transfer Delay of VTS CSMA with Clock Jitter

Simulations were performed to analyze the delay performance of VTS CSMA with different virtual time clock drift rates. Table (2) shows the delay performance with virtual time clock drift rates of 0, 10^{-4} and 10^{-3} seconds, and corresponding packet transfer delay increases of 11% and 53%, respectively.

Drift Rate [seconds]	Packet Transfer Delay [sec]	
0.0000	0.174	$T = 30$ ms $\tau = 8$ ms $\eta = 4$ $S = 0.30$
0.0001	0.193	
0.0010	0.267	

Table 2: Packet Transfer Delay of VTS CSMA with Clock Drift

Since a virtual clock jitter and drift rate of less than 0.001 seconds and 0.001 seconds/second, respectively, is easily achieved in a mobile data terminal, the performance of VTS CSMA will be relatively unaffected.

6. False Channel Status Detection

A digital busy status is transmitted on the outbound channel, by the base station, to provide the idle or busy channel information on the inbound channel. Due to channel noises, the busy status information may be decoded incorrectly by mobile terminals, and the false channel status is received. This section examines the effects of false busy/idle detection errors on the performance of VTS CSMA. Simulations were performed to analyze the delay performance of VTS CSMA with different channel status detection error probabilities, and the results are shown in table (3). The average packet transfer delay is increased by 19% with a false busy/idle detection error probability of 0.10.

For VTS CSMA in land mobile data communications, the channel status detection error probability is much less than 0.10; thus, the performance of VTS CSMA is relatively insensitive to the effects of false busy/idle detection errors when the detection error is below the above level.

channel detect. Error Prob.	Packet Transfer Delay [sec] (average)	
0.0000	0.1741	$T = 30$ ms
0.0001	0.1781	$\tau = 8$ ms
0.0010	0.1809	$\eta = 4$
0.0100	0.1837	$S = 0.30$
0.1000	0.2076	

Table 3: Packet Transfer Delay of VTS CSMA with False Busy/Idle Detection

7. Conclusion

VTS CSMA has been shown to provide high throughput, low delay performance. The results presented show that this performance is relative insensitive to the effects of packet transmission errors, channel status error, and virtual time clock jitter and drift, proving VTS CSMA to be an ideal random access protocol for land mobile data communications.

8. Reference

[1]   J.S.Meditch, D.H.-L.Yin, "Performance Analysis of Virtual Time CSMA", IEEE InfoCom, 1986, pp.242-251.

[2]   L.Kleinrock, F.A.Tobagi, "Packet Switching in Radio Channels: Part I-Carrier Sense Multiple-Access Modes and Their Throughput-Delay Characteristics", IEEE Transactions on Communications, VOL.COM-23,no.12, Dec.1975, pp.1400-1416.

[3]   M.L.Molle, D.Konstantas, "A Simulation Study of Retransmissions Strategies for the Asynchronous virtual Time CSMA Protocol", Performance 83, 1983, pp.295-308.

[4]   M.L.Molle, L.Kleinrock, "Virtual Time CSMA: Why Two Clocks Are Better than One", IEEE Transactions on Communications, VOL. COM-33. NO.9, Sept.1985, pp.919-933.

[5]   M.L.Molle, "Unifications and extensions of the Multiple Access Communication Problem", Ph.D. Dissertation, University of California, L.A., 1981, pp.77-103.

[6]   W.C.Y.Lee, "Mobile Communications Engineering", McGraw-Hill Book Company, 1982.

Analytical Performance Evaluation of the R-BTMA MAC Protocol

Sami Tabbane *, Bach Ngoc Bui **
& Philippe Godlewski ***

* MATRA-COM	** CNET	*** Télécom Paris
Rue J.P. Timbaud	Rue du Général Leclerc	Rue Barrault
78392 Bois d'Arcy	F-92131 - Issy les M.	75634 Paris

Abstract

This paper deals with an analytical modelization of the R-BTMA (Reservation- Busy-Tone Multiple Access) protocol. The R-BTMA is a MAC protocol designed for short-range Mobile Radio communications between vehicles in a decentralized context (i.e. without Base Station). The specificity of this protocol relies on the utilization of 2 different characteristic channels. Information is transmitted on the Data Channel (DC), whereas the Busy-Tone Channel (BTC) is used for signalization traffic. A cell can be seen as a group of vehicles communicating together via the DC. The range of the BTC is typically longer than the DC one. Although these physical features are conceived to globally improve the protocol performance, on the other hand, they may involve indesirable effects between neighbouring cells. The throughput/access delay performance are studied here in the particular context of two cells sharing a common BTC. The model takes into account the effect of the interactions induced by the use of the same BTC. The traffic between the vehicles is supposed to be regular and is parameter sharing : a vehicle broadcasts one packet (its "status word") per frame. The performance analysis is based on an approximate method called EPA (Equilibrium Point Analysis).

In order to improve the protocol performance, we have examined in this paper several options. First of all, the use of a signature to identify transmitted packets is introduced. The integration of a signature in each acknowledgement transmitted on the BTC allows to reduce false acknowledgement of messages. Furthermore, we have tested the waiting period adjustment before trying to access and the short lenght messages features. Integrating these 2 last items permits to improve significantly the re-access delay as well as the channel throughput. Comparison with simulation results allows us to validate the analytic model for low and medium channel offered loads.

I. INTRODUCTION

Communications in short-range mobile radio networks without base-station have to face several problems. The medium access protocol provided in such networks must take them into account.

The R-BTMA (*Reservation-Busy-Tone Multiple-Access*) protocol [13] is a merging of BTMA [6] and R-ALOHA [8] protocols. It utilizes two slotted channels (a **Data Channel** for data transmission and a **Busy-Tone Channel** for signalling) with a sliding frame structure relative to each connected station (frame synchronization is relative to each user while slot synchronization is assumed for all connected users).

The R-BTMA protocol is studied in a decentralized system without any central control and with, among others, the following problems : *hidden nodes, unforseeable and random connections, collision detection.* We observe a specific traffic nature: *parameter sharing* characteristics with *periodical transmissions*.

In this paper, we propose a performance analysis of the protocol in a 2-cells context. The system we are considering here, consists of 2 *cells* (a *cell* is seen here as a group of vehicles communicating together). We study the case in which the interference between the 2 cells gives rise to a common BTC. What is interesting for us, is to observe the influence of the BTC on the performance of the system. This is a specificity of our protocol and is one of the most difficult effect to modelize. The interactions we focuse on is that of *false acknowledgement* occurence which is a pessimistic running case for the R-BTMA protocol and the *double reservation* case (which occurs when the same time slot is reserved on the Data Channel of each cell). The performance analysis is first based on a modelization of the system behaviour by a multidimensional Markov Chain which explicitly reflects these interactions. This Markovian model is then studied by an approximate method called *Equilibrium Point Analysis* (EPA). This approach was yet considered by Tasaka in [10].

II. A LOCAL AREA NETWORK FOR SHORT-RANGE INTER-VEHICLE COMMUNICATIONS

The need for a short range inter-vehicle communication system has been pointed out by PROMETHEUS project in which our study takes place [1]. To avoid dangerous situation a vehicle must detect its neighbours and exchange information with them. The PROMETHEUS project is a *"research programme for future transportation systems with improved services to mankind, energy conservation and protection environment"* [2]. This kind of network has been considered for communications between ships within a fleet [11]. We can identify this type of Mobile Radio Networks (MRNs) by the following specific characteristics.

First of all, the most important difference between PROMETHEUS and cellular MRNs concerns the management. In PROMETHEUS, the use of Base Stations (BS) is not planned to coordinate the resources (*logical* or *physical* channels). Then, **channel management is completely distributed**. At the opposite, in cellular networks, as for instance the future GSM system [3], the BS performs channel allocation and link control.

The second characteristic on which PROMETHEUS differs from classical MRNs is the **short-range communications** between vehicles. Two reasons can be evoked to bear out this fact : a vehicle does not need to get the knowledge of a large area network topology because it is not important to receive the characteristics of vehicles hundred of meters distant. It is also quite impossible to share in real-time the characteritics of such number of vehicles with an on-board computer. The second reason is that, as frequency bandwidth must not be wasted, the system must provide an efficient dynamic frequency reuse scheme that implies a communication limited range.

Another characteristic of the PROMETHEUS system is the **localization** and **ranging** with **real-time constraints**. The system must provide at least an approximate knowledge of the LAN topology (i.e. without knowing the exact identity of each station and the whole contents of its messages). The implemented network is then to be seen more like a radar-like system to *detect, localize* and *track* other close mobile stations than like a usual LAN whose essential role is to transmit data.

The last characteristic we consider as a significant originality of the PROMETHEUS network is that connections may be established **with or without data exchanging**. The primary PROMETHEUS goal is to maintain a *logical link* with all the neighbouring vehicles. This aspect should be held as a priority with respect to the maximization of the average bit rate. This service is the very condition to insure the system safety.

Because of these specific characteristics, the functionnalities of the MAC (*Medium Access Control*) protocols in such system are very specific.

Nevertheless, noticing them, we must keep in mind that, like in other MRNs, the PROMETHEUS communications have to face important propagation problems. As the inter-mobile communications use a radio channel they have the 2 following transmission features : *near/far* or *capture effect* [4], [9], [12] and *fading*. Thus, the physical and network topology problems take an important place in the network characteristics.

Furthermore, two stations may jam each other when transmitting to a third one. Solving this problem is one of the functions of the PROMETHEUS MAC protocols.

III. The R-BTMA Protocol

Regarding to the improvements obtained by the *R-ALOHA* (Reservation-ALOHA) [5] and the *Busy-Tone Multiple-Access* [6] protocols, we have designed the R-BTMA to be suited to the previous requirements for a decentralized MRN. The R-BTMA is a protocol which insures both *contention* and *reservation*.

A. Use of two channels

We consider 2 slotted channels obtained by TDMA or FDMA (See Fig. 1):

- A DC (*Data Channel*) on which stations send their messages in the slot they had previously reserved following the R-ALOHA procedure.

- A signalling channel BTC (*Busy Tone Channel*) with a longer range than the DC one. The Busy Tone detection in a BTC slot indicates the state "busy" of the DC corresponding slot.

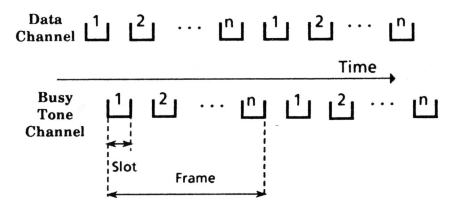

Fig. 1: Channel framing

Considering this present version of R-BTMA protocol, information transmitted on the BTC is binary:

- *Idle state* or *Collision* (Collision means that no busy tone is transmitted).

- *Busy state* with successful transmission. This case may include the **capture** occurence.

In order to improve the capture detection, each packet is marked by a **signature**. This signature can be seen as a random sequence which is included in each packet. By this mean, a user can acknowledge a well received packet by duplicating this signature into the Busy-Tone signal. When capture occurs in a slot, if the most powerful packet has a signature which is distinguishable from all

246

others, then the users whose packets have been captured are informed of their access/reservation failure. We assume that this event occurs with probability $(1 - a)$. An example of capture detection with three stations (A, B, C) is shown on Fig. 2 in which B is informed of its access/reservation failure with probability (1-a).

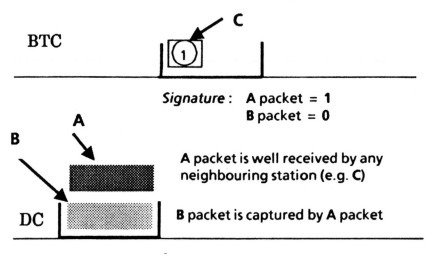

Fig. 2: Signature Mechanism

B. Protocol basic description

Before transmitting, a station (say s) is supposed to sense the 2 channels. At each moment, s notices the occupancy state of the last n slots both and on the DC and on the corresponding BTC slots. When s has a message to transmit, it chooses randomly a slot among the free ones. If number j has been chosen the station will send its packet into slot j of the next DC frame.

After the transmission of this packet, s directly switches to the receiving mode (in the receiving mode a station senses both BTC and DC). Several cases may occur:

- No Busy Tone detected or a Busy Tone with a signature different from the s one in the BTC slot j means that the packet has not been heard or understood. So, station s begins the procedure once more. In our R-BTMA protocol version, the process is repeated after a certain period of time which may be either fixed or random.

- A Busy Tone detected including the s-signature in BTC slot j indicates to station s the successful reception of its packet. Slot number j is reserved for this station until the end of emission or next lack of acknowledgement (i.e. no signal "successful" received on the BTC).

Each station detecting a correct packet in DC slot k transmits a BT including the signature of the well received packet in BTC slot k.

C. Time sharing feature

What is important in our context, is to guarantee for each user a quasi-periodic communication establishment with its neighbours in order to maintain a logical link. This requires a reduced re-access delay for each mobile station. In case of heavy message traffic load, we must control the re-access delay and the channel throughput. In order to control these parameters, we have tested the adaptative "time sharing option" (see Fig. 3).

Data Channel

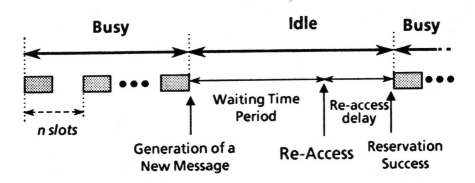

Fig. 3: Time Sharing Transmission

It consists in adapting the waiting time before trying an access regarding to the estimated offered load. This is achieved by introducing a *waiting period* without trying to access rather than being in contention and loosing slots by collision. This policy reduces the number of collisions which is the main reason of the performance degradation. Intuitively, adapting the waiting period before contention permits to have the sum *waiting time + re-access delay* lower than the *re-access delay* in the scheme without waiting. This last option will appear in our model in next section. Furthermore, time sharing is designed to be particularly well suited to short length message transmission.

D. Application of R-BTMA to PROMETHEUS

The R-BTMA, with the previous listed definition and options, has been conceived for the very specific PROMETHEUS system. Several features of this protocol meet the requirements of the future system. The distributed management feature is the essential reason of use of R-BTMA in the decentralized PROMETHEUS network. Similarly the real-time constraint is taken into account : transmitting one packet per frame per vehicle and acknowledging one slot after each well received packet, allows mobile stations to keep a good knowledge - at each moment, in real-time - of their neighbourhood topology. In addition to this, the hidden station problem is partialy solved by using the BTC : a station which detects an ACK on the BTC without having previously detected the corresponding packet on the DC can conclude to the presence of a hidden station, and be then aware. Finally, the time sharing option allows to keep acceptable performance even under high loads (which will often be encountred in PROMETHEUS) as it should be established in section V.

IV. MODEL FORMULATION

A. The model

In the following, we propose a model based on a Markovian formulation of the system behaviour. This model is analyzed by an approximate technique called *equilibrium point analysis* (EPA) [10]. The number of bursty users in each of the cells is assumed to be the same and is denoted by N.

Litterature relevant to performance analysis of MAC protocols based on only one physical channel has been extensively developed [6, 8, 9, 12]. However, as far as we know, the modelization of decentralized systems in MRNs context has not been well investigated.

The R-BTMA protocol requires the utilization of an extra channel - the BTC one - whose effects on the DC performance are difficult to modelize. Indeed, it seems illusory to set a model which contains the whole effects induced by the BTC. That is why we have chosen 2 particular schemes of running of R-BTMA. The first one is the *false acknowledgement* case which represents a gap in the protocol running. In the 2-cells context considered here, we derive the throughput/access delay performance for one cell (say the first cell) submitted to the influence of the other one (say the second cell) via the common BTC.

An example of communications between stations with *false acknowledgement* is described below.

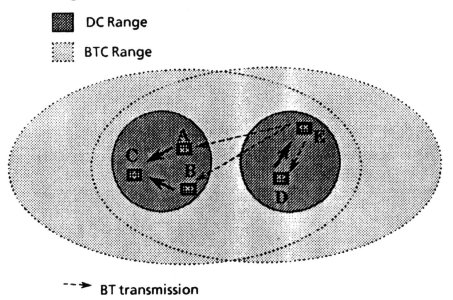

Fig. 4: The two cell context

Let's take 5 mobile stations A, B, C, D and E. A, B and C are separated from D and E by a distance greater than twice the DC range but below the BTC range. Stations A, B and D transmit a packet at the same time slot with A and B packets jamming for station C. Station E receives well D packet and acknowledges it on the BTC corresponding slot. As A and B receive this acknowledgement, they will conclude that their packets have been well received. Consequently, they will carry on their transmission and jam each other.

The second scheme we selected has good influence on the protocol performance, it is the *double reservation case*. We agree to call by *double reservation* the event whereby a user in the first cell and a user in the second cell reserve the same time slot, each one on its own DC.

We propose a model in which time is discretized at each slot. The following assumptions are adopted:

1) Users obey to the same step-transition behaviour scheme (see Fig. 5), whichever the cell they belong to. However, from one cell to the other, parameters (*probability of transmission, average number of packets per message* ...) may differ. We shall agree to distinguish the variables corresponding to the second cell by adding a dash to the notations.

2) An inactive station generates a message during a slot with probability λ. This message is further buffered before trying to access the channel. The station is then said to be in *waiting mode*.

3) A station in waiting mode decides to stop waiting with probability μ. It then tries to access the channel.

4) A user tries an access/reservation at slot t by transmitting the first packet of its message with a constant probability p_e. The slot is said to be *reserved* if a successful transmission of this packet occurs.

5) A transmitting user cannot generate a new message till his current message is not completely transmitted (see Remark).

6) Each message consists of a group of packets whose number is geometrically distributed with mean $1/\gamma$.

7) Collision and false acknowledgement detection delays are the same for all the stations and equal D slots.

8) A station which has been aware of its access/reservation failure returns in the waiting mode with probability p.

9) The probability to get a false acknowledgement is 1 - Q.

10) A double reservation occurs with probability Q_2. Conversely, a *unique reservation* occurs with probability $Q_1 = 1 - Q_2$.

Remark: The message traffic studied here takes place in the PROMETHEUS context : messages are periodically generated. Then, an active user has always a message ready-to-transmit in its buffer. This traffic model is quite different from the type of message traffic described in [8] for instance. In its model Lam assumes that messages arrival follows a Poisson law and that a user which has generated a new message immediately tries an access/reservation to the data channel without any waiting time.

B. A station behaviour modelization

Following the previous assumptions, all the users have identical behaviour. A given station may be on one of the following defined modes (see fig. 5):
T (thinking), W (waiting), A (access), K_i (*unique* reservation-i, $1 \leq i \leq n$), V_i (*double* reservation-i, $1 \leq i \leq n$), C_i (preretransmission-i, $1 \leq i \leq D$), U_i (false reservation-i, $1 \leq i \leq n$). Let us notice that we have merged T^* and T^{**} modes in one unique T mode. On figure 5, we assume $\bar{\gamma} = 1 - \gamma$.

We assume that the mode transitions of a user occur at the end of each slot. The transition scheme is reported on Fig. 5. Users having no message to transmit are said to be in T mode; they obey to assumption 2). A user in W mode obeys to assumption 3). A user in C_i mode will be ready to enter W mode after D slots with probability p. A user having reserved a slot which will appear after (i-1) slots is said to be in K_i mode if a *unique reservation* occurs, and in V_i mode if a *double reservation* occurs. We denote by R_i the sum ($V_i + K_i$). Note that the number of user in R_i mode is either *1* or *0* according to whether the corresponding slot is reserved or not. U_i mode is the analog of R_i mode with false acknowledgement occurence. In our example, a user (say **A** or **B**) which has received a false acknowledgement from **E** can detect the false reservation after a period of D slots. This occurs with probability (1-a) when **A** (or **B**) signature is different from **D** one.

C. The Analysis

The main lines of the analysis are sketched hereafter.

The first cell is described at each instant t by the state vector

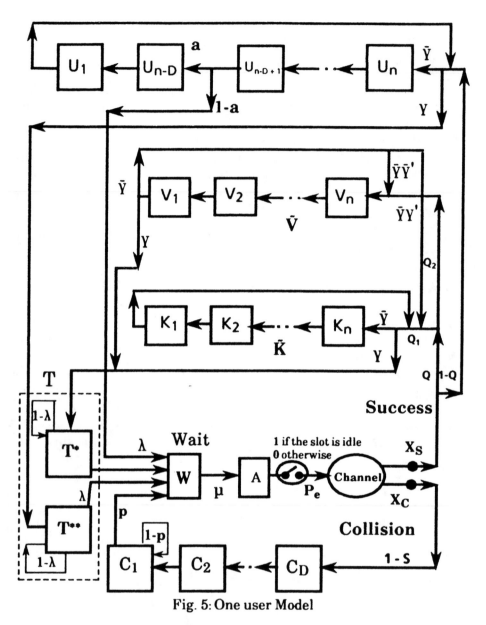

Fig. 5: One user Model

$$X(t) = (A(t), R(t), W(t), C(t), U(t))$$

where $A(t)$ (resp. $W(t)$) denotes the number of users in A (resp. W) mode at instant t; $R(t) = [R_1(t), \ldots, R_n(t)]$ is the vector whose components $R_i(t)$, $1 \leq i \leq n$, designate the number of users in R_i mode.

Similarly, we denote by C(t) and U(t) the vector corresponding to C_i mode, $1 \le i \le D$, and U_i mode, $1 \le i \le n$. Let us define B(t) = $K_1(t)$ + $K_1'(t)$ + $V_1(t)$. Note that B(t) indicates the *slot occupancy state* at instant t. If the slot is reserved (by a user which can be either in the first cell or in the second cell) then B(t) = 1, otherwise B(t) = 0. Let us introduce the conditional expectation $S_{pac}(x)$ of the number of packets transmitted without collision detection in a slot, given that the first cell is at state X(t) = x. Recall that the number of packets transmitted without collision detection includes the successful packets as well as the packets sent with false acknowledgement.

It is easily checked that X(t) is an homogenous Markov chain with finite state space. We utilize an approximate technique, called *EPA* (Equilibrium Point Analysis) to analyze this Markov chain. The principle of this analysis is based on the assumption that an equilibrium point always exists. An *equilibrium point* is defined as a point at which the expected increase in the number of users in each mode equals zero. Expressing this condition in all the modes, we can get the equilibrium point equations whose solution is the *equilibrium point* X = (A, R, W, C, U). These equations are formulated in function of:

- the **transmission, probability** that we approximate by

$$p_e = \frac{2}{n.(1-B)}$$

- the **expected number of packets** transmitted without collision detection at the equilibrium point

$$S_{pac}(X) = \frac{1}{Q}.A.p_e.(1-p_e)^{(A-1)}$$

By some algebric manipulations, this first system of equations is reduced to another one with unknowns *A, B, Q, p_e, $S_{pac}(X)$*. These unknowns are then numerically determined as the roots of the second system of non-linear equations.

We then derive the performance of the system as follow.

We define the channel throughput S(X) at equilibrium as the conditional expectation of the number of correctly transmitted packets in a slot, given that the system is at state X. Noting that *S(X) = 1* for a reserved slot and *S(X) = $S_{pac}(X)$* for a non-reserved slot, we then have:

$$S(X) = R_1(X) + \gamma.Q.S_{pac}(X).(1 - B(X))$$

We next evaluate the second performance indicator of the R-BTMA protocol: the *average access delay* D_m which is here defined to be the average time, in number of slots, from the moment a message is generated until the instant the first packet of the message is correctly received.

Applying Little's formula to the system, we deduce that

$$D_m = \frac{N}{\gamma.S} - n.(\frac{1}{\gamma} - 1)$$

where S denotes the average throughput, and $n.(1/\gamma - 1)$ is the average transmission delay of a message after a successful reservation. In our numerical results, S is approximated by $S(x)$.

V. NUMERICAL RESULTS

A. Framework of the analytical model

In our MRN, the system must run without any central management. This is an aspect which differs from protocols specified for mobile radio networks as the GSM one [3]. The model introduced here to evaluate the R-BTMA functions corresponds to a *microcell* (say the first one) interacting with another distinct microcell (say the second one) via a common BTC. The microcells communicating together via the BTC can be seen as parts of a larger cell.

Taking this feature into account, we can use the model to examine the benefits due to the use of 2 different frequency channels: the DC and the BTC.

Errors, non-received packets and *various physical layer problems* can easily be integrated and taken into account in the model.

B. Results and Discussions

In this section, numerical results are derived for the following parameters: slot number = 50 per frame, frame length = 20 ms and packet length = 320 bits. Main performance to evaluate are the *throughput* and the *re-access delay* on the DC in function of the

Channel Offered Load which is defined as the ratio: *Number of users / Number of slots per frame.*

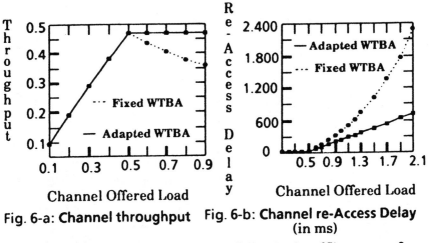

Fig. 6-a: **Channel throughput**   Fig. 6-b: **Channel re-Access Delay** (in ms)

Message length = 10 packets, Correlated Cells - Number of Signatures = 2

Fig. 6: Comparison *Fixed/ Adapted Waiting Time Before Access*

Improvements provided by adapted WTBA (Waiting Time Before Access) is shown in Fig. 6. Numerical results related to throughput and re-access delay are respectively plotted in Fig. 6-a and 6-b. On Fig. 6-a, the stabilization of the channel throughput is well reflected when choosing the option of adapted waiting time before access. Once the optimum (47%) is reached, throughput is maintained at maximum by the adapted WTBA policy for any offered higher load. Note that the rate of 47% mainly depends on message length. The other very important result (Fig. 6-b) is the decrease of the channel re-access delay. For an offered load of 200%, the delay related to the adapted WTBA policy is 680 ms while it reaches 2,000 ms for the usual policy.

Moreover, it must be noted that the shorter average message length is the better re-access delay becomes. Comparison of the curves at high offered load (see Fig. 7-a and 7-b) shows a decrease of the channel re-access delay of 75% (for a 100% offered load) against only a loss of 6% of the channel throughput in the 10 packets/message case. Choice of short messages is avantageous to reduce the channel access delay (which is a crucial parameter in our system [13]) as far as the channel throughput can be kept between acceptable bounds. Simulations and practical measurements will be the way for

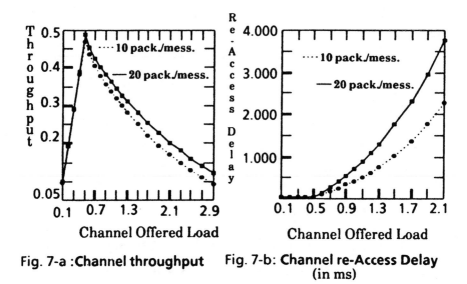

Fig. 7-a : Channel throughput Fig. 7-b: Channel re-Access Delay
(in ms)

Message length = 10 packets, Correlated Cells - Number of Signatures = 2

Fig. 7: Comparison *10/ 20 packets per message*

evaluating the optimal message length regarding to the input message traffic.

The influence of the way in which the BTC is used is shown on Fig. 8-a. We compare the DC throughput provided by only one isolated cell with that related to 2 correlated cells for a same number of users in the whole system. That is, curves of Fig. 8-a are plotted for the global system composed of two correlated cells with N users per cell and for the isolated cell system with a population of $2N$ users. Numerical results show that the maximum throughput for one cell is 76% compared to the 94% of the 2 cells context. With high channel offered load, the difference is more evident : for an offered load of 200%, the throughput one cell context is 32% and in the 2 correlated cells it reaches 64%.

We finally examine the Fig. 8-b where results from simulation and analytical methods are compared. The simulation model we use is more realistic as it takes into account some physical characterictics. At low and medium offered loads, the 2 curves are superposed, this permits to validate the results derived from simulation and analytical ways of evaluation. At high load, the results diverge in a ratio of less than a few percents. For instance, the maximum gap between

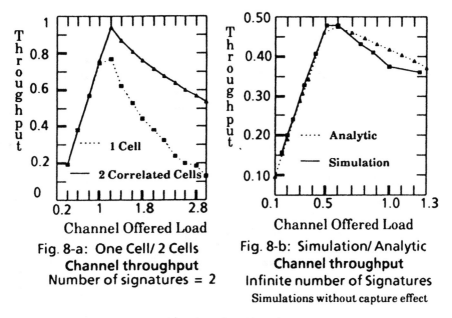

Fig. 8-a: One Cell/ 2 Cells
Channel throughput
Number of signatures = 2

Fig. 8-b: Simulation/ Analytic
Channel throughput
Infinite number of Signatures
Simulations without capture effect

Message length = 10 packets

Fig. 8: Comparison *One Cell/ 2 Cells* and *Simulation/ Analytic*

simulation an analytic values are observed for a 100% offered load; the obtained channel throughput value is 38% for simulation compared to 42% for analytical approach.

VI. CONCLUSIONS

The R-BTMA is an original protocol designed for short-range communication in Mobile Radio Networks. One specificity of this protocol is to support a decentralized control environment. Because of the complexity of the system and the physical layer characteristics, it is actually difficult to set a model taking into account all the interactions between neighbouring cells.

In this paper, we have investigated a performance analysis in a 2-cells context. More particularly, we focuse on the effect of the BTC on the *throughput/access delay* variations. The model explicitly contains the *false acknowledgement* and the *double reservation cases* as interaction effects. This model was analysed by an approximate technique called Equilibrium Point Analysis (*EPA*).

From obtained numerical results, several original conclusions can be drawn. First, a better idea about the influence of the BTC on the system performance can be set up. Computations clearly show that the use of a common BTC gives rise to 18% for the maximum channel throughput. The study of specific functions (*signature, adapted WTBA* and *short length messages*) leads us to observe some interesting effects : signature allows to reduce *false acknowledgement* occurences, adapting waiting time before access feature allows the stabilization of the throughput and a quasi-linear progression of the re-access delay instead of an exponential one. We have assumed a perfect optimization of the waiting time before access so that obtained results in that case are to be considered as optimistic. In the analytical model, we assumed that all the system parameters (*waiting users* in particular) were observable, which is not totally realistic in application. Therefore, the curves may be exploited in so far as they give a lower bound for the re-access delay and an upper bound for the channel throughput. For next studies, the model will have to take into account only the observable parameters and an estimation of the non-observable ones.

REFERENCES

[1] PRO-COM Whitebook - Aachen - August 1987.

[2] "PROMETHEUS Research Newsletter Number 5," September 1989.

[3] A. Maloberti, "Definition of the radio subsystem for the GSM public land digital mobile communications system," International Conference on Digital Land Mobile Radio Communication. Venice, June 30th - July 3rd, 1987.

[4] C. Namislo, "Analysis of mobile radio slotted ALOHA Networks," IEEE on SAC, Vol. SAC-2, no. 4, pp. 583-588, July 1984.

[5] W. Crowther, R. Retteberg, D. Walden, S. Ornstein, F. Heart, "A sytem for broadcast communication: Reservation ALOHA," Proc. 6th., HICSS, Univ. Hawaï, Honolulu, Jan. 1973.

[6] F. A. Tobagi, L. Kleinrock, "Packet Switching in Radio Channels: Part II - The Hidden Terminal Problem in Carrier Sense Multiple Access and the Busy-Tone Solution," IEEE Transactions on Communications, Vol. Com. 23, No. 12, pp. 1417-1433, December 1975.

[7] A. Murase, K. Imamura, "Idle-Signal Casting Multiple Access with Collision Detection (ICMA-CD) for Land Mobile Radio," IEEE Transactions on Vehicular Technology, Vol. VT-36, No. 2, pp. 45-50, May 1987.

[8] Simon S. Lam, "Packet Broadcast Networks - A Performance Analysis of the R-ALOHA Protocol," IEEE Trans. on Comp., Vol. C-29, No. 7, pp. 596-603, July 1980.

[9] D. J. Goodman, A. A. M. Saleh, "The Near/Far Effect in Local Area ALOHA Radio Communication," IEEE Trans. Veh. Techn., Vol. VT-36, no.1, pp. 19-22, Feb. 1987.

[10] S. Tasaka, "Stability and Performance of the R-ALOHA Packet Broadcast System," IEEE Transactions on Computers, Vol. C-32, No. 8, pp. 717-726, August 1983.

[11] A. Ephremides, "Distributed Protocols For Mobile Radio Networks," The Impact of Processing Techniques on Communications, J. K. Skwirzynski, NATO ASI Series 1985.

[12] B. Ramamurthi, A. A. M. Saleh, "Perfect-Capture ALOHA for Local Radio Communications," IEEE Journal on Selected Areas in Communications, pp. 806-813, Vol. SAC-5, No. 5, June 1987.

[13] S. TABBANE, P. GODLEWSKI, "R-BTMA : a MAC Protocol for Short-Range Mobile Radio Communications," Proceedings of 1990 IEEE VT Conference, Orlando, pp. 582-587, May 1990.

Multiple Access Technique for Radio-Local Area Networks.

Z. IOANNOU, M. K. GURCAN, and H. C. TAN

Digital Communications Section, Department of Electrical Engineering,
Imperial College of Science, Technology and Medicine, London, U.K.

Abstract- *In this paper a multiple access technique is considered. This technique is assigned to operate in accordance with a proposed Radio-Local Area Network (R-LAN), in which radio channels are statistically distributed among cells. In studying this technique, we consider packet-switched data traffic only and assume queued rather than a blocking system. To assess the performance of this network under varying traffic conditions, a multiple access model, which is employed for the traffic modelling of a cell, is developed. Under the assumption of Poisson message arrivals, the generating function of the system size is obtained and from that the mean packet delay is derived.*

1. Introduction.

The need to provide flexible and cost-effective communications within a limited area and particularly in large office buildings, initiated the development of R-LANs, radio local area networks. As opposed to fixed-wire local area networks (F-LANs), this technology provides wireless access to a variety of telecommunications networks.

The basic considerations in the design of R-LANs are similar to cellular radio systems. The coverage area, say a large office building, is divided into very small cells. In each cell a multichannel base station provides service to users belonging to the coverage area of that cell. The same frequency channels can be reused in cells which are situated far apart, above or below.

Although the design principles of R-LANs are similar to cellular radio, as yet, there are no existing standards to represent their technology. The most notable work towards this trend is the development of DECT (Digital European Cordless Telecommunications) which is currently in the design stages [AKER88]. It will operate using a Time Division Multiple Access together with Time Division Duplex (TDMA/TDD), i.e. all channels will be paired together on the same carrier for uplink and down link transmission.

Recently, various access schemes which are assigned to operate in a local wireless environment are attracting considerable attention. In [GOOD89], packet reservation multiple access (PRMA), a technique for transmitting, over short range radio channels, a mixture of voice and packets from various information sources, is proposed and studied. An integrated voice/data system is proposed in [ZHAN90] based on the idea of movable boundary TDMA system. A framed polling technique is used as the channel access protocol. Here, we describe a multiple access model which is a reservation scheme.

2. Description of the multiple access model.

In the proposed R-LAN, a base station in a cell is allocated a set of radio channels with statistically distributed channel availabilities. Each channel independently alternates between 'on-periods' and 'off-periods' referred to as *service* and *blocked* periods, respectively. The allocation of those radio channels is based on a novel resource allocation method, with signal to interference ratio (S/I) and bit error rates (BERs) as prime criteria [TAN90].

The network is periodically updated. Depending on the traffic demand in a cell, more or less channels or indeed longer or shorter service periods are allocated accordingly. Thus, the overloading of the system is prevented and this procedure provides a form of flow control. Within an updating time interval the service and blocked periods of each channel are controlled by a periodic process with fixed period. However, if the availability of each channel is observed over a large number of updating intervals this process is no longer deterministic, but is approximated by a random distribution.

A star architecture is assumed for the R-LAN. Thus, each user in a cell communicates directly with a base station using a multiple access scheme (see Fig. 1). Within the coverage area of a cell, there are users (wireless terminals) scattered around and each user has a source of information, a local queue and a low power transmitter. A source of information of a user generates messages which consist of a number of fixed-length data packets. However, before transmitting those packets an access reservation packet is transmitted to the base station for a reservation for the desired number of packets. Upon successful transmission of the reservation packet, the actual data message to be transmitted is queued up waiting for the appropriate reserved data slot or slots over which transmission can take place. Thus, each reservation is for a position in the common queue of the base station for a single or a group of data packets. Using the broadcast capability of the base station, a reservation packet successfully transmitted is received by all users and each user maintains a reservation queue counter. Therefore, it is sufficient for each user to know the queue length (in number of packets) as well as the queue positions of his own reservations. The queue discipline is first-in-first-out (FIFO) according to the order reservation are received.

In order to examine the behaviour of the system in terms of the traffic demand (arrival rate) and the available channels within a cell, we shall assume that after a successful reservation a user places his packets in a common queue (the queue at the base) referred to as *the base station queue* (BSQ). Each packet is then taken out of the queue and is transmitted independently in one of the available channels of the *statistically distributed slot assignment* (SDSA) scheme. Thus, the access scenario which is illustrated in Fig. 2 consists of two stages; the reservation stage and the transmission stage.

We are mainly concerned with the uplink transmission (user-to-base station) and the emphasis is given on the time delay a packet spends in the transmission stage of the system. Here, we derive the system size in the transmission stage (that is, the number of packets in the system seen by a successful reservation), when the radio channels in the statistically distributed slot assignment (SDSA) scheme are randomly distributed. This corresponds to the case where the base station in each cell is allocated a set of radio channels, and the availability of each channel is independently governed by a random distribution. In this study, we examine the case where this random distribution is described by Bernoulli process.

Therefore, the transmission stage is modelled as a multichannel queueing system with randomly distributed channel availabilities, in which the packet arrival process is the assumed output of the reservation stage. The channels are slotted in time and each time-slot, which is of fixed length, can support the transmission of a single data packet. Without loss of generality, we can let this transmission time equal unity. Further, data packets are synchronously transmitted; that is, each data packet is taken out synchronously from the queue (BSQ) for transmission at each time-slot (discrete clock time). The data packets arriving at the BSQ during a time-slot have to wait to begin transmission at the beginning of the next time-slot, even if the queue is empty at the time of arrival.

The above system can be thought of as a queueing model with gates between the servers and waiting room which are randomly and independently opened at the beginnings of time-slots, regardless of whether or not the waiting area is empty. Therefore, the transmission stage of our system may be represented by the model of Fig. 3.

3. The generating function of the system size.

For the purpose of the analysis it is assumed that the capacity of the global queue is infinite (the data traffic is delayed rather than blocked) and the queue discipline is FIFO. We also assume that each message consists of a single packet (i.e. simple arrival process) and the transmission is error-free (i.e. no retransmissions). Each packet is independently transmitted whenever there is a channel (time-slot) available. The packet arrivals at the base station are assumed to be independent and identically distributed (i.i.d.) random variables. The aggregate packet arrival is Poisson distributed with mean arrival rate λ packets/slot. Similarly, service and blocked periods of the channels are i.i.d. random variables. Furthermore, in this study we assume that each channel is independently controlled by a Bernoulli process. Thus, channel i is in a *service* state with probability p_i and in a *blocked* state with probability $1-p_i$.

The system is examined just after the beginning of a time-slot (beginning of a service interval) and packets arriving within a time-slot are considered for transmission at the beginning of the next time-slot. Packet transmission can start only at the beginning of a time-slot.

It is noted that the structure of the system is completely synchronous and a discrete-time analysis is desirable. For this analysis, let us also consider the following random variables:

Q_k : denotes the number of packets in the system (including those in service) just after the beginning of the kth time-slot. This can be seen to be the number of packets in the queue at the end of time-slot.

G_k : number of packet arrivals during the kth time slot.

c : number of channels in the system.

S_k : number of available channels (slots) during the kth time-slot.

Given the number of packets in the system at the beginning of a particular time-slot, the number of packets in the system at future time-slots depends only on the new arrivals and the available channels but not on past system size. Thus, we have an imbedded Markov chain similar to the classical $M/G/1$ queueing system [KLEI75].

The equilibrium system size is described by the probability generating function (p.g.f.), denoted by $Q(z)$, where

$$Q(z) = E[z^Q] = lim_{k \to \infty} Q_k(z) \qquad (1)$$

and $E[x]$ denotes the expected value of x. Under the assumption of Poisson arrivals and Bernoulli distributed channel availabilities the p.g.f. of the system size, obtained in [IOAN90], is given by

$$Q(z) = \frac{\left\{(c-\lambda) - \sum_{i=1}^{c}(1-p_i)\right\} \cdot (z-1)}{z^c \cdot e^{-\lambda(z-1)} - \prod_{i=1}^{c}\left\{(1-p_i) \cdot z + p_i\right\}} \cdot \prod_{r=1}^{c-1} \frac{z - z_r}{1 - z_r} \qquad (2)$$

where z_r, $r=1, 2, \ldots, c-1$, are the $c-1$ zeros of the denominator inside the unit circle $|z_r| < 1$. Note that on setting all p_i's equal to unity (i.e. all channels are continuously available) we obtain the well known p.g.f. of the system size for the $M/D/c$ queueing system with unit service time [PRAB65].

4. Mean packet delay.

In order to find the mean system size we make use of the moment generating property of the p.g.f.. Thus, the mean system size just after the beginning of a time-slot is given by

$$\overline{q} = \frac{d\,Q(z)}{dz}\bigg|_{z=1} \qquad (3)$$

Equation (3) expresses the mean system size at the beginning of a time-slot. However, since the packets can only enter service at the beginning of a time-slot, whereas packet arrivals can occur continuously, we need to add the approximate time average number of packet arrivals during the service interval (time-slot), which is $\lambda/2$ for Poisson arrivals [CHU70]. This has the effect of adding a delay component of 1/2 slot which is true if the actual packet arrivals are uniformly distributed over each time-slot. Therefore, the mean system size is expressed as

$$\overline{n} = \lambda/2 + \overline{q} \qquad (4)$$

Taking the first derivative of (2) and applying *L'Hopital's* rule (twice) for the limit $z=1$, the mean system size is given as

$$\overline{n} = \frac{\lambda}{2} + \frac{2\lambda c - \lambda^2 - c(c-1) + \frac{d^2}{dz^2}\left\{\prod_{i=1}^{c}\left\{(1-p_i)z + p_i\right\}\right\}\bigg|_{z=1}}{2 \cdot \left\{(c-\lambda) - \sum_{i=1}^{c}(1-p_i)\right\}} + \sum_{r=1}^{c-1} \frac{1}{1 - z_r} \qquad (5)$$

Using *Little's* formula [KLEI75], the mean packet delay in units of time-slots is given by

$$\overline{d} = \overline{n}/\lambda \qquad (6)$$

In addition, we would like to define a parameter for the channel utilisation, ρ. For a steady state to exist the mean packet arrivals in a time-slot must be less than the mean number of channels (slots) available in that time-slot. Thus, the channel utilisation, which is always less than unity, must satisfy the condition

$$\rho = \lambda/\overline{S} < 1 \qquad (7)$$

where $\overline{S}$ represents the mean number of the available channels (slots) in a time-slot. The mean number of the available channels (slots) in a time-slot, $\overline{S}$, can be found using the moment property of the p.g.f. of $s(r)$. Denoting $S(z)$ to be the p.g.f. of the probability distribution of channel availabilities, then it's mean value is given by the first moment evaluated at $z=1$, $S'(1)$. Thus, the channel utilisation described in (7) may be rewritten as

$$\rho = \lambda/S'(1) \qquad (8)$$

5. Numerical results.

In order to obtain any numerical results, one must find the c zeros (roots) of the denominator of (2) inside the unit circle. To obtain, however, all the zeros of the denominator of (2) in one sweep is proving to be a difficult task. Although standard iterative methods may be used for such computation, their effectiveness is based on the supposition that one knows the approximate location of those roots in the first place in order to initiate iterations. For illustration purpose, and without loss of generality, we consider a system with two channels only.

We now look at different examples. In these examples, the availability of the first channel is described by the probability p_1, whereas the second channels is available with probability p_2. Both probabilities are taken to be equal and different set of results have been obtained for various values of p_1 and p_2. In Fig. 4 the mean packet delay is plotted against the mean packet arrival rate for the case of $p_1=p_2=0.1$, 0.2, 0.3 0.4, 0.5, 0.6, 0.7, 0.8, 0.9, and 1. As it can be noted, when the channels have higher availabilities (i.e. higher probabilities) the system can support greater arrival rates, thus, higher traffic demand. This is very evident from the performance of the discrete-time $M/D/2$ system where the channels are continuously available which, of course, corresponds to $p_1=p_2=1$.

The mean packet delay grows without bound as the mean arrival rate approaches a given value which is readily noted from (5) to be $\lambda=(p_1+p_2)$ packets/slot. On the average, there is a delay component of 1.5 time-slots which includes 1 time-slot of transmission time and 1/2 time-slot due to the fact that arriving packets in a time-slot are only considered for transmission in the next time-slot. This discrete-time system exhibits on the average 1/2 time-slot higher packet delays than a conventional continuous-time system.

Also, for a system with two channels, the channel utilisation as described in (8) becomes

$$\rho = \lambda/S'(1) = \lambda/(p_1+p_2) \tag{9}$$

In Fig. 5 the channel utilisation is plotted against the mean packet delay for each example, using (9). It is noted that high channel utilisations can be achieved, however, at the expense of higher packet delays.

6. Conclusions.

We have described and analysed a multiple access technique which is assigned to operate in accordance with a R-LAN in which the channel availabilities are statistically distributed among cells. The results indicate that it is possible to achieve good throughput delay performance.

Although a system with continuous availability of channels may support higher traffic demands, it is evident that a system with controllable channel availabilities provides a higher degree of flexibility. This flexibility is achieved by varying the availabilities of the channels in order to meet changing traffic demands. In doing so, the system becomes adaptive to traffic variations. The channels can be shared among base stations and the system capacity may be utilised more efficiently, however, at the expense of slightly higher packet delays.

Perhaps the main advantage of the proposed R-LAN lies in the area of flow control. That is, because the network is periodically updated, flow throughout that network is entirely controllable. Therefore, congestion control is easily facilitated since the base station which undergoes traffic congestion is easily pointed out and the effect of congestion is confined.

In this study, we have concentrated on the transmission stage of the access model and confined ourselves to data traffic only. However, in order to obtain the overall packet delay encountered in our system, it is necessary to consider the effect of the reservation stage. Also, in a packet communication environment, the integration of voice and data is very important. These issues are currently being investigated.

References

[AKER88] D. Akerberg, 'Properties of a TDMA picko cellular office communication system', *GLOBECOM'88*, 1988, pp. 1343-1349.

[CHU70] W. W. Chu, 'Buffer behavior for Poisson arrivals and multiple synchronous constant outputs', *IEEE Trans. Comput.*, Vol. C-19, No. 6, June 1970, pp. 530-534.

[GOOD89] D. J. Goodman, et. al., 'Packet Reservation Multiple Access for Local Wireless Communications', *IEEE Trans. Commun.*, Vol. 37, No. 8, 1989, pp. 885-890.

[IOAN90] Z. Ioannou, *Multiplexing and Multiple Access Techniques for Radio-Local Area Networks*, M.Phil./Ph.D. Transfer Report, Digital Communications Section, Depart. Elect. Eng., Imperial College of Science, Technology and Medicine, London, August 1990.

[KLEI75] L. Kleinrock, *Queueing Systems, Vol. I: Theory*, Wiley, New York, 1975.

[PRAB65] N. U. Prabhu, *Queues and Inventories*, Wiley, New York, 1965.

[TAN90] H. C. Tan, *Dynamic Resource Allocation for Radio-Local Area Networks,* M.Phil./Ph.D. Transfer Report, Digital Communications Section, Depart. Elect. Eng., Imperial College, London, August 1990. Part of this report also to appear in this workshop.

[ZHAN90] K. Zhang and K. Pahlavan, 'An integrated voice/data system for mobile indoor radio networks', *IEEE Trans. Veh. Techn.*, Vol. 39, No.1, 1990, pp. 75-82.

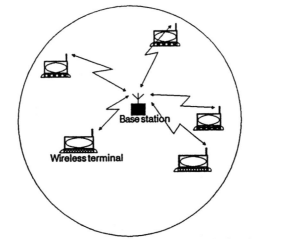

Fig. 1.   Channel access scenario in a cell.

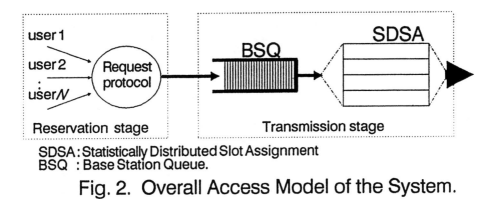

SDSA : Statistically Distributed Slot Assignment
BSQ : Base Station Queue.

Fig. 2. Overall Access Model of the System.

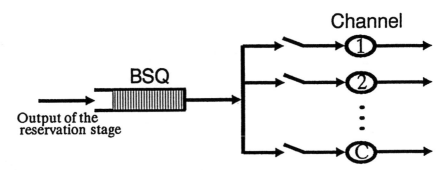

Fig. 3. Queueing model of transmission stage.

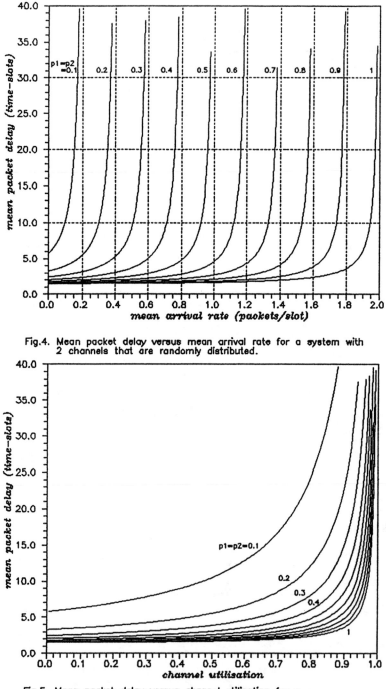

Fig.4. Mean packet delay versus mean arrival rate for a system with
2 channels that are randomly distributed.

Fig.5. Mean packet delay versus channel utilisation for a
system with 2 channels that are randomly distributed.

Spread Spectrum Wireless Information Networks for the Small Office

Raymond W. Simpson
O'Neill Communications, Inc.
100 Thanet Circle
Princeton, NJ 08540

1. Introduction

Most of the personal computers in the US are in small businesses and are not networked. Small businesses have special needs which haven't been fully addressed by traditional LAN systems. This paper will explore these requirements and show how wireless information network products address some of these needs, with examples based on the OCI LAWN®. In addition, newer wireless network systems which will address the small business are discussed.

2. Requirements of the Small Office Environment

In this context, the term "small office" means primarily the small business office, which must operate profitably on its own, and does not have a larger corporate umbrella of support services available to it. In some larger organizations, individual activities may fit this model, especially field offices, individual offices of franchise chains and the like.

The primary considerations for the small office are a high level of perceived usefulness and a low cost of ownership. High perceived usefulness means it solves problems the users recognize or gives them the ability to do things they can't easily do now.

2.1 Cost of Ownership

The cost of ownership includes: the cost to purchase, cost to install, cost to learn to use, cost of disruption to the business while learning the new system, cost to maintain, reconfigure and expand the system. Wireless systems certainly reduce the cost of installation, expansion and reconfiguration. If the system operates under Part 15 of the FCC Rules, there is no cost of obtaining a license or service provider fees further reducing cost. The cost of learning to use the system and the disruption to the business is not primarily a function of the wireless aspect (except there is no physical disruption needed to install cables), but is mainly a function of the supporting software of the network system.

The purchase cost is perhaps the most difficult part of wireless networks, as a network device which includes a radio transceiver is inevitably greater than the cost of one employing only a wire transceiver, assuming equivalent levels of integration and mass manufacture. Presently, neither of these two assumptions hold, but at least the gap in volume and level of integration may be driven closer by the growth of short range wireless system such as CT-2, DECT and PCN. The increased cost is traded against the cost of wiring which can be far greater, with estimates running as high as $1000 per node. However, these estimates are unrealistic in most small office situations -- there is usually a low cost substitute, such as telephone wiring, the power wiring or some semi-technical employee with a roll of twisted pair.

2.2 Perceived Usefulness

Small office users are usually recognize their own inertia -- if they think it will be hard to learn to use a system they know they won't use it and they won't buy it. This is also part of the cost of ownership discussed above. Most small networks are used primarily as peripheral sharers, with some electronic mail and file activities. If the potential user can see how to use it for sharing peripherals right away, there is a good chance of a sale.

Transmission speed in networks is a major factor in network engineering. Speed is important in the small office environment, but speed is measured by how quickly the user can continue with his work, not in over the air data rate. This is a fortunate outcome for developers of low cost wireless networks operating in limited available RF bandwidth. The appearance of fast operation can be had by making printing or electronic mail delivery a background task. Remember, even the best LAN can't get the paper output of the printer back to the user's desk, it is still necessary to get up and go get it.
The increased popularity of graphics printing will increase the required minimum bit rate, but most graphic print data contain long runs and can be highly compressed for transmission. Also, the conversion of graphic document to printable form is a fairly long process on most personal computers, so a moderate bit rate on the wireless link has a minor effect on the time from pressing the print key to having a paper document to review. Printing graphic documents in uncompressed form at 20 kb/s in background mode has proved to be acceptable in day to day operations.[1]

The exception to the above is file operations, especially if a file needs to be retrieved before work can proceed. File transfer in a background mode is acceptable for archiving and as a means of non-paper document routing, however. The high data rates of modern wire and fiber networks is more a function of two factors: the need to serve a large number of users in larger businesses and the lesser bandwidth restrictions of wire and fiber.

3. Wireless Link Requirements and Part 15 Operation

The link requirements are considered here in radio terms, although many of these same factors also apply to free space optical systems. The main link requirements are:

Direct Range: enough to cover a small office, typically 30 m is adequate
Throughput:
    for peripheral sharing: 20 kb/s to 50 kb/s
    for file sharing: >100 kb/s
Reliability: links that work should work (almost) always

The particular combination of requirements makes the small office wireless network a prime candidate for operation under Part 15, especially the spread spectrum provisions. It also leads to spread spectrum designs significantly different from the designs used in more traditional spread spectrum systems.

Transmission impairments affecting system design, especially in the Part 15 spread spectrum band include:

    multipath
    shadowing
    interference
    effect of link type on access protocol

Spread spectrum is helpful in reducing the effects of multipath propagation, but cost limitations reduce the options available to the system designer (RAKE receivers are still cost prohibitive for the small office). Switched diversity, on the other hand, is low in cost if only one receiver is used.

Shadowing in the office environment is caused by metallic partitions, metal studs in plaster walls, wire lath in walls of older buildings, masonry walls and, especially, concrete and steel floor construction. The designer has few tools at his disposal against shadowing, but digital store-and-forward repeating is one of the most cost effective (when the reduction in throughput is acceptable).

Part 15 confers no protection whatsoever against interference, but the office environment does. Generally, the office has control over its own premises, so interfering sources can be excluded from the premises (but not emissions from off-premises emitters). The attenuation with distance characteristics at UHF in buildings have been characterized in various ways[2345] but all agree the attenuation within buildings is quite high except in the case where the propagation is in unobstructed open spaces. This works against long RF links within buildings, but also works against interference sources from outside the premises. The signal from a low power LAWN® within a small office is almost always the dominant signal in the 902-928 MHz. band even when other Part 15 devices (including LAWNs) are in use in other offices in the same building.

The combination of premises control with the spread spectrum processing gain (which is often much less than in military system for both cost and occupied bandwidth reasons) and other standard techniques such as frequency agility and packet retries can provide a robust system in most office environments. In LAWN, a 16 chips per bit spreading sequence in combination with a selection of 4 subbands or channels allows multiple independent networks and frequency agility to avoid interference.

The 4 channels also allow users to change frequency in the unlikely event their LAWN system should cause interference to a licensed service. Licensed activity in 902-928 MHz includes government, automatic vehicle monitoring and amateur radio[6]. This aspect of operation under Part 15 is often neglected in the rush for higher bit rates. A Part 15 system which uses the whole 902-928 MHz band for one direct sequence signal, operating at the full 1 watt level, will radiate between 0.5 and 1 mW in the input bandwidth of a typical narrowband FM repeater. This is sufficient to produce harmful interference a kilometer away. Even worse, not all the communication systems in the band are narrowband FM, some AVM systems are wideban
d and operate with very sensitive receivers[7]
A full-band system has no freedom to change frequency to avoid causing the interference -- in such a situation, the user of the system is legally obligated to cease the operation which causes the harmful interference <u>even if it means shutting down his system.</u>[8]

4. Near Term Future

There are a number of established wireless network products operating at higher speeds, including the Telesystems Arlan product line[9] and the Agilis system[10]. These are currently in the over $1000 per node price class. NCR has recently announced a 2 Mb/s system[11], and OCI has a high speed direct sequence system under development and a 2.45 GHz integrated voice/data system in laboratory prototype form.

A major theme for these higher speed networks will be compatibility, in the sense of easy inter-operation, with established network standards. This is a goal of the IEEE P802.11 working group. Some degree of standardization has been started with the above mentioned systems, which each feature operation with a major network software system. This promises the purchaser a growth path and freedom from obsolescence. It doesn't yet provide easy setup, administration and training for the small business.

The ideal small business system will need a "beginners network system" which also provides:

1. Transparent use of peripherals
2. Fast, easy installation and setup
3. Usable by partially computer literate office workers with no more than one hour of study or training
4. Simple, effective file access control / security system which takes no more than 1 hour to set up and is almost maintenance free.

Wireless connectivity is clearly part of the solution -- installation is very fast. But significant simplifications need to be made in other areas to properly penetrate the small business environment. Integrated wireless voice/data systems can provide the small business solution. Not only is digital cordless telephony provided, but voice printing and voice recognition can be incorporated to allow voice command telephone control and user authentication for security management. This would provide small business security on the level most natural for a small business: authority and privilege are granted to persons not to machines.

Notes

1. Printer Sharing Devices, Infoworld, Sept. 10, 1990, p.82.

2. Theodore S. Rappaport, Factory Radio Communications, RF Design, July 1989, p 67.

3. T. S. Rappaport and C.D. McGillem, UHF Multipath and Propagation Measurements in Manufacturing Environments,IEEE Globecom 1988 Conference Record, p. 26.5.1

4. T. Koryu Ishii, RF Propagation in Buildings, RF Design, July 1989, p. 45.

5. D. Akerberg, Properties of a TDMA Pico Cellular Office Communication System, IEEE Globecom 1988 Conference Record, p. 41.4.1

6. Title 47 Code of Federal Regulations, Parts 2, 90, 97

7. Bill Goshay, Pactel Teletrac, personal communication,1990

8. Title 47 Code of Federal Regulations, Part 15

9. Telesystems SLW, Inc., 85 Scarsdale Road, Suite 201, Don Mills, Ontario M3B 2R2.

10. Agilis Corporation, 1101 San Antonio Road, Mountain View CA 94043.

11. InfoWorld, Sept. 3, 1990, p. 25.

AN OVERVIEW OF CODE DIVISION MULTIPLE ACCESS (CDMA) APPLIED TO THE DESIGN OF PERSONAL COMMUNICATIONS NETWORKS

Allen Salmasi

Vice President - Planning & Development
General Manager - Digital Cellular/PCN
QUALCOMM Incorporated
10555 Sorrento Valley Road
San Diego, CA 92121-1617

1.   Introduction and Summary

CDMA is a modulation and multiple access scheme based on spread spectrum communications techniques, a well established technology that has been applied only recently to digital cellular radio communications and advanced wireless technologies such as PCN [1]-[2]. It solves the near-term capacity concerns of major markets and answers the industry's long-term need for a next generation technology for truly portable communications in the most economic and efficient manner providing for a graceful evolution into the future generations of wireless technologies.

In the U.S., it is now widely believed that the first generation "Advanced Wireless Technologies" and "Personal Communications Network (PCN)" services will be provided by the cellular carriers using the next generation digital cellular system infrastructure and the cellular frequencies as a straightforward extension of the cellular system. A breadboard CDMA digital cellular was system developed by QUALCOMM and demonstrated in San Diego, CA in conjunction with PacTel Cellular to over 150 representatives of the cellular industry in November 1989. In February 1990, an extensive technical field trial was carried out in midtown Manhattan with NYNEX Mobile. For the purposes of carrying out the CDMA trials and demonstrations a total of 3 cell-site and 2 mobile prototype units were built. In each case, the demonstration system utilized two actual cell-sites and one mobile, and included A/B comparisons with the commercial analog system. The fundamental features of CDMA such as power control and soft hand-off were tested and proven in a variety of field conditions.

Arrangements with Other Equipment Suppliers

To provide for large scale deployments of the CDMA network and subscriber equipment on a worldwide basis in 1992, QUALCOMM has reached technology licensing agreements with two of the largest network equipment suppliers in the world, AT&T and MOTOROLA, for joint development and manufacture of CDMA digital network and subscriber equipment. QUALCOMM has also made agreements with one other major network vendor and four of the largest subscriber equipment suppliers in the world. As of November 1990, AT&T, MOTOROLA, Nokia and Clarion have made public announcements. Most importantly, cellular companies representing at least seven of the top ten largest cellular markets and numerous other top 100 markets are now seriously considering CDMA digital cellular and its PCN extension as their primary choice for the provision of the next generation of advanced wireless communications services in the United States.

Highlights of Current Development Activities

On July 31, 1990, QUALCOMM released the first draft of the "CDMA Digital Common Air Interface (CAI) Standard, Cellular System Dual-Mode Mobile Unit - Base Station Compatibility Standard", to a number of major cellular carriers and network as well as subscriber equipment manufacturers that are actively involved in the development of the CDMA digital cellular equipment. Based on the comments received from the participants, Revision 1.0 of the CAI, which is a significant interim release, was released on October 1, 1990. Most significantly, the fundamental system design approach was field tested both before and during the development of the CDMA CAI specification. Over the next several months, the most significant activities are:

1) Completion of the design and the development of the mobile unit and the base station channel unit chip sets. There are a total of five Application Specific Integrated Circuits (ASICs) that perform all of the digital signal processing required by the mobile unit (three ASICs) and the channel unit (three ASICs); one ASIC is common to both the mobile unit and the channel unit;

2) At present, a large number of "brassboard" multi-sector cell sites and mobile units are being manufactured for "real world environment" testing in the Summer of 1991. Initial brassboard

equipment will be used for comprehensive system tests and for tests of the ASICs prior to and after their release to fabrication. First pass ASICs will then be used to expand the scope of the tests to include the Field Validation of the CDMA CAI Specification and a comprehensive field test of CDMA system capacity under "fully loaded with real interferers" channel conditions, with a large number of cell sites and mobile equipment. Release of the validated Revision 2.0 of the CAI is scheduled for the Fall of 1991, which will allow for full scale manufacturing of operational equipment. The remainder of 1991 will be used for "full-up" system testing of production equipment, including full MTSO and network functions, with infrastructure equipment manufacturers.

The Standards Issue

On December 12, 1988 the Federal Communications Commission issued a Report and Order in General Docket 87-390. 3 FCC Rcd 7033 (1988). In the Report and Order the Commission amended the rules governing the use of the cellular spectrum to "provide a regulatory structure that (would) facilitate the implementation of a new generation of cellular technology and services . . . ." *Id*. at 7033. In particular, the Commission gave cellular licensees the "option to use (a) portion of their spectrum to implement advanced cellular technologies or auxiliary common carrier services, provided several important conditions are satisfied." *Id*. These conditions included "limitations on the signal strength at the cellular service area boundaries to ensure against interference" and "requiring a complete technical analysis of any potential interference before implementing advanced technology cellular or auxiliary service." *Id*.

Therefore, there is absolutely no requirement imposed on cellular carriers and/or equipment manufacturers by the U.S. Government to follow the TIA IS-54 Standards, or any other proposed standards, in developing and implementing the next generation digital cellular system. The standards that will be considered "the U.S. Standards" are the *de facto* standards set by the cellular carriers and manufacturers in ordering, developing and implementing operational systems in the United States.

The following is a list of the major attributes of the CDMA digital wireless telephony system developed by QUALCOMM:

Capacity -- The capacity limits of CDMA have now been extensively analyzed and tested to the possible extent and are now projected to be in the range of 15 - 20 times that of existing FM analog technology as a result of the properties of a wideband CDMA system such as improved coding gain/modulation density, voice activity gating, 3-sector sectorization and reuse of the same spectrum in every cell. The capacity of CDMA will double with the application of the next generation half-rate vocoders.

Availability -- With the full development of production mobile cellular radios and cell site equipment now underway, CDMA can be ready for system implementation by early 1992.

Economies -- CDMA is a cost-effective technology that requires fewer, less expensive cells and no costly frequency re-use patterning, offers inexpensive mobile and portable radios, and provides for much lower system cost per subscriber because of large capacity.

Call Quality -- CDMA takes advantage of multipath, enhancing performance in urban areas and virtually eliminating multipath fading and static; its soft hand-off feature and soft capacity limit reduces the number of lost and blocked calls.

Portability - The high capacity, and low transmitter power and moderate signal processing power requirements for the portable terminals (compared to the other approaches), low cost of implementing and maintaining PCN type of services as an extension of the existing cellular infrastructure, and the ease of operations encourages growth to a seamless, integrated Personal Communications Network.

Voice Quality -- CDMA transmission of voice by a high bit rate vocoder ensures superior, realistic voice quality. CDMA provides for variable data rates allowing many different grades of voice quality to be offered which can be priced accordingly.

Privacy -- The "scrambled signal" of CDMA completely eliminates cross-talk and makes it very difficult and costly to eavesdrop or track calls, ensuring greater privacy for callers and greater immunity from air-time fraud.

Transition - Because CDMA provides large performance improvements, the capacity of even the first digital systems deployed using only 10% of the current analog allocation is more than sufficient to provide for a low probability of blocking with a fairly significant population of digital subscriber units operating in the field.

2. What is CDMA?

The multiple access problem can be thought of as a filtering problem. We have a multiplicity of simultaneous users desiring to use the same electromagnetic spectrum, and an array of filtering and processing techniques from which to choose to allow the different signals to be separately received and demodulated without excessive mutual interference. The techniques that have been devised include the use of propagation mode selection, spatial filtering with directive antennas, bandpass filters, and time sharing of a common channel. More recently, techniques involving spread spectrum modulation have evolved in which more complex waveforms and filtering processes are employed.

Spatial filtering uses the properties of directive antenna arrays to maximize response in the direction of desired signals and to minimize response in the direction of interfering signals. The current analog cellular system uses sectorization to good advantage to reduce interference from co-channel users in nearby cells.

Each CDMA signal consists of a different pseudo-random binary sequence which modulates the carrier, spreading the spectrum of the waveform. A large number of CDMA signals share the same frequency spectrum. If one looks at CDMA in either the frequency or the time domain, the multiple access signals appear to be on top of each other. The signals are separated in the receivers by using a "correlator" which accepts only signal energy from the selected binary sequence and despreads its spectrum. The other users' signals, whose codes do not match, are not despread in bandwidth and as a result, contribute only to the noise. The signal-to-interference ratio is determined by the ratio of desired signal power to the sum of the power of the other signals enhanced by the system "processing gain", the ratio of spread bandwidth to baseband data rate. If all the signals arrive at the same power level then the interference-to-signal ratio is simply approximately equal to the number of signals, or the capacity.

A CDMA system can also be a hybrid of FDMA and CDMA techniques where the total system bandwidth is divided into a set of wideband channels, each of which contains a large number of CDMA signals. Hybrids of TDMA and CDMA are also possible.

QUALCOMM has devised a novel system approach to the design of cellular radio networks using CDMA. In this approach, an orthogonal spectrum spreading code structure is utilized to allow a large number of terminals to share the same spectrum. In QUALCOMM's design of the

282

CDMA digital cellular system for U.S., the multiple access scheme exploits isolation provided by the antenna system, geometric spacing, power gating of transmissions by the voice activity signal, careful mobile and cell-site transmit power control, and very efficient modem and signal design using error correction coding.

The CDMA digital cellular waveform design uses a Pseudo-Random Noise (PN) spread spectrum carrier. The chip rate of the PN spreading sequence is chosen so that the resulting bandwidth is about 1.23 MHz after filtering, is approximately one-tenth of the total bandwidth allocated to one cellular service carrier. The Federal Communications Commission (FCC) has allocated a total of 25 MHz for mobile-to-cell-site and 25 MHz for cell-site-to-mobile for the provision of cellular services. The FCC has divided this allocation equally between two service providers, the Block A and the Block B operators, in each service area.

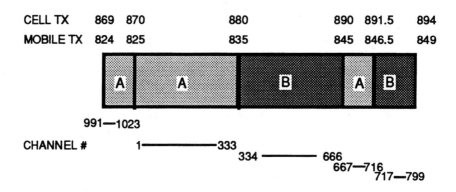

FIGURE 2.1 CELLULAR FREQUENCY ALLOCATIONS

The channel number denotes the FM channels (30 kHz channels). Because of the order in which allocations were made, the 12.5 MHz allocated to each carrier for each direction of the link is further subdivided into two sub-bands. For the wireline carriers, the sub-bands are 10 MHz and 2.5 MHz each. For the non-wireline carriers, the sub-bands are 11 MHz and 1.5 MHz each. A signal bandwidth of less than 1.5 MHz could be fit into any of the sub-bands, while a bandwidth of less than 2.5 MHz could be fit into all but one sub-band.

Thus, in order to preserve maximum flexibility in matching the CDMA technique to the available frequency spectrum, the waveform must be less than 1.5 MHz in bandwidth. A good second choice would be a bandwidth of about 2.5 MHz, allowing full flexibility to the wireline carriers and nearly full flexibility to non-wireline carriers in the U.S. Going to a wider bandwidth offers the advantage of increased multipath discrimination but with the disadvantage of increased cost of the equipment and lower flexibility in frequency assignment within the allocated bandwidth.

A set of ten 1.23 MHz bandwidth CDMA channels would be used if the entire allocation were converted over to CDMA. In the interim, only one or a small number of 1.23 MHz channels need to be removed from the present FM analog service to provide digital service. Each such 1.23 MHz segment can provide over twice the capacity of the entire 12.5 MHz allocation using the present FM system. ·

2.1. The CDMA Signal and Waveform Design

The following section is excerpted from the CDMA Digital CAI Standard entitled "Cellular System Dual-Mode Mobile Station-Base Station Compatibility Standard," Draft Revision 1.0, dated October 1, 1990, produced by QUALCOMM, Inc. This document specifies the common air interface for a system that uses CDMA to provide a very high capacity digital cellular telephone system. All data and information contained in or disclosed in this section are proprietary information of QUALCOMM, Inc. and all rights therein are expressly reserved.

2.1.1. CDMA Forward Link Waveform Design

The CDMA CAI specifies a forward link CDMA waveform design that uses a combination of frequency division, pseudo-random code division and orthogonal signal multiple access techniques. Frequency division is employed by dividing the available cellular spectrum into nominal 1.23 MHz bandwidth channels. Normally, a cellular system would be implemented in a service area within a single radio channel until demand requires employment of additional channels.

Pseudo-random noise (PN) binary codes are used to distinguish signals received at a mobile station from different base stations. All CDMA signals in the system share a quadrature pair of PN codes. Signals from different cells and sectors are distinguished by time offsets from the basic code. This relies on the property of PN codes that the auto-correlation, when averaged over a few bit times, averages to zero for all time offsets greater than a single code chip time (approximately 1 μsec).

The PN codes used are generated by linear shift registers that produce a code with a period of 32768 chips. The PN chip rate is 1.2288 MHz, or exactly 128 times the 9600 bps information transmission rate. Two codes are generated, one for each of two quadrature carriers, resulting in quadriphase PN modulation. To avoid confusion between the system bandwidth, the PN chip rate, and the frequency assignment spacing, note that the PN chip rate is exactly 1.2288 MHz. The frequency assignment spacing, in multiples of 30 kHz, for two adjacent CDMA carriers is 1.23 MHz. Note that the 3 dB bandwidth is also 1.23 MHz.

The signals are bandlimited by a digital filter that provides a very sharp frequency roll-off, resulting in a nearly square spectral shape that is 1.23 MHz wide at the 3 dB point. Signals transmitted from a single antenna in a particular CDMA radio channel share a common PN code phase. They are distinguished at the mobile station receiver by using a binary orthogonal code based on Walsh functions (also known as Hadamard matrices). The Walsh function is 64 PN code chips long providing 64 different orthogonal codes. Orthogonality provides nearly perfect isolation between the multiple signals transmitted by the base station.

The information to be transmitted is convolutionally encoded to provide the capability of error detection and correction at the receiver. The code used has a constraint length (encoder memory) of nine, K=9, and a code rate of one-half (two encoded binary symbols are produced per information bit), r=1/2. To provide communications privacy, each data channel is scrambled with a user addressed long code PN sequence. Thus, a "channel" in the forward link of the specified CDMA system consists of a signal centered on an assigned radio channel frequency, quadriphase modulated by a pair of PN codes with an assigned time offset, biphase modulated by an assigned orthogonal Walsh function, and biphase modulated by the encoded, and scrambled digital information signal.

An important aspect of the forward link waveform design is the use of the pilot signal that is transmitted by each cell-site and is used as a coherent carrier reference for demodulation by all mobile receivers. The pilot is transmitted at a relatively higher level than other types of signals allowing extremely accurate tracking. The Pilot Channel signal is unmodulated by information and uses the zero Walsh function (which consists of 64 zeroes). Thus, the signal simply consists of the quadrature pair PN codes. The mobile station can obtain synchronization with the nearest base station without prior knowledge of the identity of the base station by searching out the entire length of the PN code. The strongest signal's time offset corresponds to time offset of the nearest base station's PN code. After synchronization, the pilot signal is used as a coherent carrier phase reference for demodulation of the other signals from this base station. Figure 2.2 shows an example of all of the signals transmitted by a base station on a particular sector antenna. Out of the 63 forward code channels available for use, the example shown in the figure depicts seven paging channels (the maximum number allowed) and 55 traffic channels. Other possible configurations could replace the paging channels one for one with traffic channels, up to a maximum of no paging channels and 63 traffic channels.

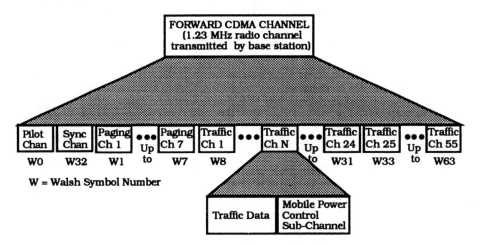

FIGURE 2.2 EXAMPLE OF LOGICAL FORWARD CDMA CHANNELS TRANSMITTED BY A BASE STATION

The remaining synchronization details and other system information are communicated to the mobile station by the base station's synchronization channel. This channel has a fixed assigned Walsh function. Once the sync channel has been received, the mobile station can select one of the paging channels to listen for other system information and pages.

2.1.2. CDMA Reverse Link Waveform Design

The CDMA reverse link also employs PN modulation using the same 32768 length binary sequences that are used for the forward link. Here, however, a fixed code phase offset is used. Signals from different mobile stations are distinguished by the use of a very long (2^{42} - 1) PN sequence with a user address determined time offset. Because every possible time offset is a valid address, an extremely large address space is provided. This also inherently provides a reasonably high level of privacy.

The transmitted digital information is convolutionally encoded using a rate 1/3 code (three encoded binary symbols per information bit) of constraint length nine. The encoded information is grouped in six symbol groups (or code words). These code words are used to select one of 64 different orthogonal Walsh functions for transmission. The Walsh function chips are combined with the long and short PN codes. Note that this use of the Walsh function is different than on the forward link. On the forward link, the Walsh function is determined by the mobile station's assigned channel while on the reverse link the Walsh function is determined by the information being transmitted.

The use of the Walsh function modulation on the reverse link is a simple method of obtaining 64-ary modulation with coherence over two information bit times. This is the best way of providing a high quality link in the fading channel with low Eb/No where a pilot phase reference channel cannot be provided. Note that on the forward channel the Pilot Channel signal is shared among all the users of the forward channel. This is not possible on the reverse channel.

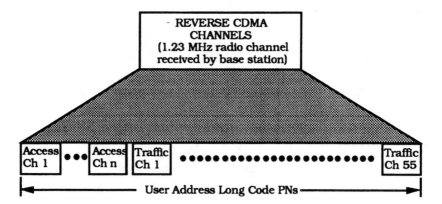

FIGURE 2.3. EXAMPLE OF LOGICAL REVERSE CDMA CHANNELS RECEIVED AT A BASE STATION

A "channel" on the reverse link of the specified CDMA system consists of a signal centered on an assigned radio channel frequency, offset quadriphase modulated by a pair of PN codes, biphase modulated by a long PN code with address determined code phase, and biphase modulated by the Walsh encoded and convolutionally encoded digital information signal.

Figure 2.3 shows an example of all of the signals received by a base station on a particular sector antenna. Each Reverse CDMA Channel can have up to 62 Traffic Channels and up to 32 Access Channels per supported Paging Channel.

3. The CDMA System Design

In a co-channel interference limited environment such as cellular telephone, CDMA provides much better frequency reuse.

- In CDMA, the same spectrum is reused in every single cell of the system
- CDMA has much lower C/N than FDMA or TDMA
- CDMA can exploit voice activity for higher capacity
- CDMA provides sectorization capacity gain nearly equal to the number of sectors
- CDMA requires lower fade margins to operate in difficult mobile Rayleigh faded environments

3.1. CDMA For Greatly Increased Capacity

In the cellular frequency reuse concept, interference is accepted but controlled with the goal of increasing system capacity. CDMA does this more effectively because it is inherently a better anti-interference waveform than FDMA or TDMA. Indeed, its genesis was in military anti-jamming systems. Narrowband analog modulations are limited in frequency reuse efficiency by the requirement to achieve a Carrier-to-Interference ratio of about 18 dB. This requires that a channel used in one cell not be reused in a nearby cell. In CDMA, the wideband channel is reused in every cell. A combination of open loop and closed loop power control (through measurements of the received power at the mobile station and the base station) is used to command the mobile station to make power adjustments so as to maintain a power level for adequate performance. This minimizes interference to other users, helps to overcome fading, and conserves battery power in the mobile station.

In CDMA, frequency reuse efficiency is determined by the signal-to-interference ratio resulting from all the system users within range, not just those in any given cell. Since the total capacity becomes quite large, the statistics of the users are what is important, not any single user. The "law of large numbers" can be said to apply. This means that the net interference to any given signal is simply the average of all the users' received power times the number of users. As long as the ratio of received signal power to the average noise power density is greater than a threshold value then the channel will provide an acceptable signal quality. With TDMA and FDMA, interference is governed by a "law of small numbers" in which "worst-case" situations determine the percentage of time in which the desired signal quality will not be achieved.

The primary parameters that determine the CDMA digital cellular system capacity are processing gain, the E_b/N_0 (with the required margin for fading), the voice duty cycle, the CDMA omni-directional frequency reuse efficiency, and the number of sectors in the cell-site antenna. Additionally, for a given blocking probability, the larger number of voice channels provided by CDMA results in a significant increase in trunking efficiency, serving a larger number of subscribers per voice circuit.

To develop the equation for capacity we start with the equation for C/I noting that C, the received signal power from a mobile at the cell is, by definition, equal to $R \cdot E_b$ where R is the transmission bit rate and Eb is the signal energy per bit and I is, by definition, equal to $W \cdot N_0$, where W is the system transmission bandwidth, and No is the interference power spectral density. The C/I ratio is therefore:

$$\frac{C}{I} = \frac{R \cdot E_b}{W \cdot N_o}$$

Eb/No is defined as the ratio of energy per bit to the noise power spectral density that is required by the particular modulation and coding scheme utilized. In the above equation, W/R is commonly called the processing gain of the system.

In a multiple access system using spread spectrum (CDMA), we apply the above equation by noticing that the interference power, I, is equal to the $C \cdot (N-1)$ where N is the number of users transmitting in the band, W. Thus, C/I is equal to 1/(N-1). This assumes that all the signals are transmitted at a controlled power level so as to arrive at the receiver with power, C. This results in equation (1) below for the capacity of CDMA in a non-cellular environment with no additional system features.

$$N - 1 = \frac{W}{R} \cdot \frac{1}{\dfrac{E_b}{N_o}} \qquad (1)$$

For example, if a spread spectrum bandwidth of 1.25 MHz is utilized by mobiles transmitting at 9600 bits per second and if the modulation and coding technique utilized requires an E_b/N_o of 6 dB, then up to 32 mobiles could transmit simultaneously provided that they are each power controlled to provide equal received power at the receiving location. In a cellular system, this capacity is reduced by interference received from neighboring cells and increased by other factors, as discussed in more detail below.

Because mobile terminals in a cell are at different ranges and experience different path losses, power control of the inbound signals is necessary to normalize the power received at the cell-site from all the mobiles operating in its cell. Each mobile measures the power level received from the cell-site and estimates the path loss. The larger the power received by a mobile from the cell-site, the smaller the path loss to the cell-site and, therefore, the smaller the transmit power required in the inbound path. This power control process will eliminate received power variations due to differing mobile to cell range, differing terrain, etc. It will not eliminate Rayleigh fading because the phase relationships that cause this kind of fading are not correlated over the 45 MHz frequency difference between the outbound and inbound frequencies in the U.S. cellular system.

The high-speed closed loop power control measures the power received from each mobile at the cell-site receivers and commands a power adjustment to normalize all received signals within the cell. The QUALCOMM system is capable of compensating for differing multipath between the two directions of the link. The combination of the open loop and closed loop power control techniques results in a very wide dynamic range, very high speed power control that compensates for all known effects. The demonstration system has proven to perform as predicted.

3.2. Achievement of Low E_b/N_0

E_b/N_0 is the ratio of energy per bit to the noise power spectral density and is the standard figure-of-merit by which digital modulation and coding schemes are compared. It is directly analogous to C/N for analog FM modulation. With the CDMA system it is possible to use extremely powerful, high redundancy error correction coding techniques due to the wide channel bandwidth employed. With narrowband digital modulation techniques, a much higher E_b/N_0 is required compared to CDMA because less powerful, low redundancy error correction codes must be used to conserve channel bandwidth. QUALCOMM has employed a powerful combination of forward error correction coding together with an extremely efficient digital demodulator in its implementation of the CDMA digital cellular system.

As mentioned earlier, in the cell-site to mobile direction, or forward link, the CDMA signal design uses convolutional encoding with constraint length K=9 and code rate 1/2. The optimum decoder for this type of code is the soft decision Viterbi algorithm. A standard design is used for this purpose.

In the mobile to cell-site link, or reverse link, since the modulation scheme employs 64-ary orthogonal signalling based on the set of Walsh function sequences, the demodulation is based on the use of the Fast Hadamard Transformers (FHT) as an optimum receive filter for the Walsh function. In the cell-site receiver, the correlator output is fed to a FHT processor. This processor produces a set of 64 coefficients for every 6 symbols. The 64 symbols are then multiplied by a weighting function and passed to a diversity combiner. The weighted 64 symbols from each antenna's receiver are added to each other. The resulting set of 64 coefficients is tested to determine the largest coefficient. The magnitude of the result, together with the identity of the largest of the 64 is used to determine a set of decoder weights and symbols for a Viterbi algorithm decoder. The Viterbi decoder, a constraint length K=9, code rate 1/3 decoder, determines the most likely information bit sequence. For each vocoder data block, nominally 20 msec of data, a signal quality estimate is obtained and transmitted along with the data. The quality estimate is the average signal-to-noise ratio over the frame interval.

3.2.1. Multipath Performance of CDMA

The CDMA system will offer much improved performance in the urban environment because CDMA provides inherent discrimination against multipath. Any delay spread greater than 1 microsecond will be uncorrelated and appear only as weak additional interference. This feature of CDMA is a very important attribute, because it eliminates the need for a costly and power consuming equalizers which has its most profound impact on the design of portables and pocket phones, especially for the PCN application.

The cellular mobile channel typically consists of the Rayleigh faded components without a direct line-of-sight component. The Rayleigh fading channel is caused by the signal being reflected from many different features of the physical environment resulting in copies of a signal arriving simultaneously from many directions with different transmission delays. At the UHF frequency employed for mobile radio communications, this causes significant phase differences between the paths with the possibility for destructive summation of the signals which in turn results in deep fades. This class of fading is known as frequency-selective or time-flat fading and effects only a part of the signal bandwidth. The fading is a very strong function of physical position of the mobile unit. A small change in position, even a few inches, changes the physical delay associated with all of the paths resulting in a different phase for each path. Thus, the motion of the mobile through the environment can result in a quite rapid fading process. In the 800 MHz cellular radio band, this fading is typically as fast as one fade per second per mile per hour of vehicle speed. Fading is very disruptive to the channel and results in poor communication quality and requires additional transmitter power to overcome the fades.

The path loss of the UHF mobile telephone channel can be characterized by two separate phenomena; an average path loss which can be described statistically by a log-normal distribution whose mean is proportional to the inverse fourth-power of the path distance and whose standard deviation is approximately equal to 8 dB. The second phenomena is a fading process caused by multi-path which is characterized by a Rayleigh distribution. The log-normal distribution can be considered to be the same for both the inbound and the outbound frequency bands (as used in conventional UHF cellular telephone systems). However, the Rayleigh fading is an independent process for the inbound and outbound frequency bands. The log-normal distribution is a relatively slowly varying function of position. The Rayleigh distribution, on the other hand, is a relatively fast varying function of position.

The wide bandwidth PN modulation allows different propagation paths to be separated when the difference in the path delays for the various paths exceeds the PN chip duration (or 1/bandwidth). This is because the PN sequence has zero correlation for time offsets greater than one chip time. For a PN chip rate of 1.2288 MHz as employed in the present design approach CDMA system, the full spread spectrum processing gain, equal to the ratio of spread bandwidth to system data rate, can be employed to discriminate against paths that differ by more than one microsecond. A one microsecond path delay differential corresponds to differential path distance of 1000 feet. The urban environment typically provides differential path delays ranging from less than one microsecond up to 10-20 microseconds in some areas.

In relatively narrowband modulation systems such as analog FM modulation employed by the first generation cellular phone system, the existence of multiple paths causes severe fading. With wideband CDMA modulations, however, the different paths may be discriminated against in the demodulation process, greatly reducing the severity of the multipath fading. Multipath fading is not completely eliminated because there will occasionally exist two or more paths with delay differentials of less than one microsecond (in a 1 MHz PN chip rate system) which cannot be discriminated against in the demodulator. This will result in fading behavior.

Diversity is the favored approach to mitigate fading. There are three major types of diversity: time diversity, frequency diversity and space diversity. Time diversity can best be obtained by the use of interleaving and error detection and correction coding. Time diversity is usually helpful only when the channel conditions are time varying. Wideband CDMA offers a form of frequency diversity by spreading the signal energy over a wide bandwidth. Space or path diversity is obtained by providing multiple signal paths through simultaneous links from the mobile to two or more cell-sites, by exploitation of the multipath environment through spread spectrum processing, allowing signals arriving with different propagation delays to be received separately, and by providing multiple antennas at the cell-site (and in some cases, in the mobile). Time diversity and antenna diversity can easily be provided in FDMA and TDMA systems. However, the other methods can only be provided easily with CDMA. The greater the order of diversity in a system, the better will be the performance in this difficult propagation environment.

A unique capability of direct sequence CDMA is the exploitation of multipath to provide path diversity. If two or more paths exist with greater than 1 microsecond differential path delay, then multiple PN receivers can be employed to separately receive the strongest of signals in multiple paths. The number of signals (or paths) is equal to the number of PN receivers provided. Since these signals will typically exhibit independence in fading, i.e., they usually do not fade together, the outputs of the receivers can be diversity-combined so that a loss in performance only occurs when all receivers experience fades at the same time. This type of diversity not only mitigates Rayleigh fading but also fades caused by blocking of the signal path by physical obstructions. Signals arriving with larger than 1 microsecond delay spread will likely also arrive from different directions. Signals coming from different directions will be affected differently by obstructions in the immediate physical environment of the mobile. This effect is especially pronounced in heavily built-up urban areas.

3.3. Voice Activity Detection

It has been shown that in a typical full duplex two-way voice conversation the duty cycle of each voice is, on the average, less than 35%. It is not cost-effective to exploit the voice activity factor in either FDMA or TDMA systems because of the time delay associated with reassigning the channel resource during the speech pauses. With CDMA, it is possible to reduce the transmission rate when there is no speech, substantially reducing interference to other users. Since the level of other user interference directly determines capacity, the capacity is increased by nearly a factor of two. This also reduces mobile transmit requirements by a factor of nearly two. If we define the transmit duty cycle as d, then the interference power received is now equal to $N \cdot d$. Equation (1) above then becomes:

$$N - 1 = \frac{W}{R} \cdot \frac{1}{\frac{E_b}{N_o}} \cdot \frac{1}{d} \qquad (2)$$

3.4. Frequency Reuse in CDMA Cellular Systems

The fundamental advance of analog cellular radio systems over its predecessors is the concept of frequency reuse. Analog FM voice modulation requires 18 dB C/I isolation to provide acceptable performance. Digital FDMA and TDMA modulation techniques require similar isolation. The employment of frequency reuse techniques in cellular radio systems has allowed much higher capacity than previous mobile telephone systems, resulting in the current boom in demand for cellular radio service.

In equations (1) and (2) above, only the interference from mobile units within the boundaries of a cell was considered. We now consider the case of a large cellular system and calculate the interference received in a cell from mobile units operating in neighboring cells. Let us assume a large number of equal sized cells, and uniform density of mobile terminals. If the system existed in an area of uniform flat topography with relatively low antennas then the path loss is well known to follow a fourth power law of distance. This is fortunate (for FDMA and TDMA also) because without this path loss, an unacceptable level of interference would be received from far distant stations in a very large area system.

The analysis for an FDMA cellular system is based on the interference that comes not from the statistical average of received interference from all other users in the system, but rather from particular units in the nearby cells using the same frequency. The interfering cell must be far enough away so that the received interference power will result in a C/I of greater than 18 dB. Simple geometric arguments[1] for the 7 cell cluster frequency reuse case with omni antennas show that for mobiles positioned in the worst case locations within the two cells an average C/I of 22.3 dB results. However, the log-normal distribution of path loss will result in C/I less than 18 dB much of the time. In fact the C/I exceeded 90% of the time is only about 16 dB. This illustrates the need for greater than 7 frequency sets when omni antennas are used or else the use of directive antennas to provide adequate isolation. Use of 120° sectors improves the C/I situation by about 6 dB for FDMA.

With 7 cell cluster frequency reuse, FDMA/FM can provide about $\frac{12.5 \times 10^6}{30 \times 10^3 \cdot 7}$ or about 57 simultaneous calls per cell (and 2 control channels) in a 25 MHz system bandwidth (inbound and outbound).

In CDMA, the total interference at the cell site to a given inbound mobile signal is comprised of interference from other mobiles in the same cell plus interference from mobiles in neighboring cells. The contribution of all the neighbor cells is equal to approximately half the interference due to the mobiles within the cell. The frequency reuse efficiency of omni-directional cells is shown below to be simply the ratio of interference from mobiles within a cell to the total interference from all cells, or about 64%. Figure 4.1 shows Percentage of Interference Contributions From Neighboring Cells.

In the following we derive the above result. First, let us assume only spatial isolation effects in the cellular system, i.e., we will consider that the cell-sites have omni-directional antennas. The results will then be augmented to include the effect of directive cell-site antennas.

In a cell containing N mobile transmitters, the number of effective interferers will simply be N-1 regardless of how they are distributed within the cell since power control is employed in the mobiles. The

---

[1] W.C.Y. Lee, Mobile Cellular Telecommunications Systems. McGraw-Hill Book Company, 1989.

power control operates such that the incident power at the center of the cell from each mobile will be the same as for every other mobile in the cell, regardless of distance from the center of the cell. As discussed above, a novel combination of open loop and closed loop power control (through measurements of the received power at the mobile and the cell-site) is used to command the mobile to make any necessary adjustments to maintain the desired power level. This conserves battery power in the mobiles, minimizes interference to other users and helps to overcome fading.

In a hexagonal cell structure, six cells are immediate neighbors of the central cell. The mobile units within these neighboring cells will control their power relative to their own cell center. Let us assume that the path losses for mobiles in the neighboring cells to their own cell centers is also fourth-law. Of course, the interference path loss from adjacent cell mobiles into the center cell is also fourth-law.

The total signal to interference ratio received at a cell-site is:

$$\frac{C}{I} = \frac{1}{N + 6Nk1 + 12Nk2 + 18Nk3 + ...} \quad (3)$$

which can be factored to yield:

$$\frac{C}{I} = \frac{1}{N [1 + 6k1 + 12k2 + 18k3 + ...]} \quad (4)$$

where N is the number of mobiles per cell, and k_i, i=1, 2, 3, ... are the interference contribution from individual cells in rings 1, 2, 3, etc., relative to the interference from the center cell. This loss contribution is a function of both the path loss to the center cell and the power reduction due to power control to an interfering mobile's own cell center.

Let us define a frequency reuse efficiency, F, as:

$$F = \frac{1}{1 + 6k1 + 12k2 + 18k3 + ...}$$

Numerical integration techniques and/or simulation techniques can be used to calculate F. The result is that the frequency reuse efficiency, F, is about 0.65 for this propagation model. Figure 3.1 below shows the relative interference contributions of each cell surrounding the center cell.

A similar set of simulations can be used to show that the outbound half of the CDMA system's capacity is comparable to the inbound or reverse channel. Various software simulations and field tests have verified these results.

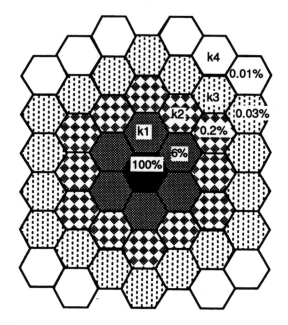

FIGURE 3.1 - PERCENTAGE OF INTERFERENCE CONTRIBUTIONS FROM NEIGHBORING CELLS

3.5. Sectorization Capacity Gain

When directional cell-site antennas are used, the typical 120° sector antennas, for example, the interference seen is simply divided by three because, on the average, it only looks in the direction of 1/3 of the mobiles. The capacity supportable by the total system is therefore increased by a factor of three.

3.6. The Complete CDMA Capacity Equation

The complete equation for determination of capacity in QUALCOMM's CDMA system given by equation (1) augmented by the additional system capabilities to provide the following equation for CDMA capacity:

$$N = \frac{W}{R} \cdot \frac{1}{\dfrac{E_b}{N_o}} \cdot \frac{1}{d} \cdot F \cdot G \qquad (5)$$

Where:

N = Calls Per Cell •••••••• Assuming: Rayleigh Fading, Reverse Link
W = Spread Spectrum Bandwidth •••••••••••••• Assuming: 1.25 MHz
R = Data Rate in kbps •••••••••••••••••••••••• Assuming: 9600 bps
E_b/N_o = Bit Energy + Noise Power Spectral Density • Assuming: 6.0 dB
D = Voice Duty Cycle •••••••••••••••••••••••••• Assuming: 50%
F = Frequency Reuse Efficiency •••••••••••••••••• Assuming: 60%
G = Number of Sectors in Cell ••••••••••• Assuming: 3 sectors (120°)
Radio Capacity Per Cell •••••••• 120 CDMA Channels in 1.25 MHz
Erlang Capacity Per Cell (2% Blocking) •••• 107 Erlangs in 1.25 MHz

The Erlang capacity calculation was made using the pessimistic assumption that the soft capacity limit feature described below is not used; that is, that when the 41st call comes to a particular sector that it is refused service and does not try again. The soft capacity limit could allow such a call to proceed with the understanding that slightly higher bit error rates would result for all system users.

4. Personal Communications Systems

Although QUALCOMM's first terrestrial application of the CDMA is for the development of the next generation digital cellular system, it is important to realize that the CDMA technology being developed, and the chip-set "engines", could be cost-effectively used in many other important applications. The major wireless personal communications applications are:

- CDMA for Microcells,

- Public Cordless Systems (e. g., Telepoint)

- Wireless Local Loop, and

- Wireless PBX,

Current and future cellular radio systems must be able to operate in an environment of increasing spectrum congestion and interference levels. One efficient way of using the available spectrum is to reduce the cell size and tailor its shape to the teletraffic demands. This dramatically increases the capacity of the system, reduces the radiated power levels (a biological plus) and reduces the equipment complexity. There is no universally accepted definition of a microcellular system. For some, it is a conventional cellular system with smaller cells, designed mainly for areas with dense teletraffic demands. Microcells may be classified as either one-, two-, or three-dimensional, depending on whether they are along a road or a highway, covering an area such as a number of adjacent roads, or located in multilevel buildings, respectively. The number of microcells per cluster may be much higher than with current systems.

Systems may operate independently, in conjunction with a PBX, or the public switched telephone network (PSTN); handsets would be small, lightweight and usable in multiple environments, while digital technology would enhance speech quality and allow a greater degree of security than is afforded by currently used handsets. Recent studies conducted at QUALCOMM demonstrated that with transmit power levels of less than 18 milliwatts in the CDMA mobile, it is possible to provide coverage to street blocks of 500 meter in length. Similar power level and coverage results are obtained for in-building (3-dimensional) and hot spots (2-dimensional) microcells. The key advantages of CDMA are:

- Much higher capacity than other techniques
- Rugged links in multipath and Rayleigh fading
- Soft handoff
- Flexible application: adjacent cells and sectors may be as different as the capacity distribution dictates
- Lower portable transmit power means longer battery life
- Competitive cost with other digital technologies

[1] Allen Salmasi, " Future Public Land Mobile Telecommunication Systems (FPLMTS) Employing Code Division Multiple Access (CDMA) and Advance Modulation and Coding Techniques," QUALCOMM Information Paper to the Sixth meeting of CCIR IWP 8/13, Harrogate, U.K., July 1990.

[2] K.S. Gilhousen, I. M. Jacobs, R. Padovani, A. J. Viterbi, L. A. Weaver, and C. E. Wheatley, "On the Capacity of a Cellular CDMA System," to appear in IEEE Transactions on Vehicular Technology, May 1991.

CDMA POWER CONTROL FOR WIRELESS NETWORKS

Jack M. Holtzman
Rutgers University, WINLAB
Dept. of Electrical and Computer Engineering
P.O. Box 909, Piscataway, New Jersey 08855-0909

ABSTRACT

Code division multiple access (CDMA) has been proposed for wireless networks including digital cellular. Power control is an important aspect of such systems. We consider here power control in the reverse link direction. The objective is to equalize the powers received at a base station from different mobiles. Otherwise, those users with low received power could have severely degraded performance. We address two questions: How well need it be done? and How well can it be done? For the first question, we quantify the loss in capacity due to imperfect power control. For the second question, we discuss the response speed needed for a feedback controller to combat Rayleigh fading.

1. INTRODUCTION

Direct sequence code division multiple access (DS/CDMA) has been proposed for wireless networks including digital cellular. It has been recognized that power control is important for its performance. The reverse link (up-link) problem is specifically addressed here. (The forward link problem is a different issue; see [1] for that and other issues). In the reverse link, the powers received at the base station from the different users should not be too different to prevent the low power users (low received powers) from suffering severely degraded service. The issue is pressing as capacity calculated from the reverse link direction would be expected to be more limiting than in the forward direction (the calculations of [1] show that).

We address two questions here:

1. How well need power control in the reverse link direction be done?

2. How well can it be done?

For the first question, we specifically consider how much capacity (in number of users for a given bit error probability) must be reduced to compensate for imperfect power control. The analysis uses recent results [2] which facilitate taking effects such as unequal power into account. The results are easy to use, accurate, and permit easy calculation of key quantities. We consider the first question for the uncoded case and for convolutional coding. For the uncoded case, the results of [2], extended for unequal powers, are used. For convolutional coding, an upper bound on bit error probability is used.

For the second question, we discuss the response speed requirements for the effectiveness of feedback control in combatting Rayleigh fading.

Other work on DS/CDMA with unequal power includes [3]-[5]. In [3], results on unequal power are given for some specific examples. Refs. 4 and 5 analyze the effect of unequal power due to multipath fading. Our orientation is somewhat different in that a "reasonably effective" power control is assumed to keep the received powers "almost" equal and we want to see the implications of them not being exactly equal.

Section 2 describes the system and relevant results from [2]. These results are extended to the case of unequal powers in Section 3. Section 4 gives numerical results for the model of Section 3, i.e., for the uncoded case. In Section 5, results for convolutional coding are given. Section 6 addresses the second question on how well can power control be done. Concluding remarks are given in Section 7.

2. SYSTEM MODEL

The system model description follows that of [6] which, in turn, uses results of [7]. The system is phase coded with the data signal modulated onto a phase-coded carrier. There are K users with the kth user's transmitted signal being of the form[1]

$$s_k(t-\tau_k) = \sqrt{2P}\, b_k(t-\tau_k) a_k(t-\tau_k) cos(\omega_c t+\phi_k) \qquad (2\text{-}1)$$

where $b_k(t)$ and $a_k(t)$ are the data and spreading signals, respectively, P is

---

1. For the relationship of the phase-coding to this equation, see [8].

the received signal power, ω_c the carrier frequency, and ϕ_k the phase. The data signal $b_k(t)$ is a sequence of unit amplitude (positive and negative) rectangular pulses. The spreading signal $a_k(t)$ is a faster sequence of unit amplitude (positive and negative) rectangular pulses (chips). There are N chips per data pulse. τ_k and ϕ_k are time and phase shifts. The pulse and chip amplitudes are all independent, identically distributed random variables with probability of 1/2 of being ±1. During demodulation at the receiver, the composite of all the users' signals is multiplied by a synchronized replica of the original signature sequence. Note, in particular, that all of the users' received powers (P) are equal in the above. In Section 3, that is relaxed.

The decision statistic (after correlation) for the desired signal 1, normalized with respect to the chip duration T_c and with all signals' received power $P=2$, is

$$Z_1 = N + \sum_{k=2}^{K} W_k \cos \Phi_k \qquad (2\text{-}2)$$

The sum in (2-2) is the multiple-access interference (MAI). W_k is given by

$$W_k = P_k S_k + Q_k(1-S_k) + X_k + Y_k(1-2S_k) \qquad (2\text{-}3)$$

Relative to user 1, the kth interfering user has a time offset to the nearest chip given by S_k and carrier phase Φ_k (i.e., $S_k = \tau_k - \tau_1$). The descriptions of P_k, Q_k, X_k, and Y_k are given in [2] and [6] and will not be repeated here. Another key quantity is B (in the equations below) which represents the number of chip boundaries in the desired signal 1 at which a transition to a different value occurs.

An improved Gaussian approximation is given in [6] which is based on the observation that the MAI is approximately Gaussian, *conditioned* on the delays and phases of all the interfering signals and on B. Then an accurate approximation to the bit error probability is given by

$$\hat{P}_e = \int_0^\infty Q\left[\frac{N}{\sqrt{\Psi}}\right] f_\Psi(z)\,dz \qquad (2\text{-}4)$$

where $Q(x)$ is a cumulative normal integral from x to ∞ and

$$\Psi \equiv Var[MAI \mid S, \Phi, B] \qquad (2\text{-}5)$$

with $S = (S_2, \cdots, S_K)$ and $\Phi = (\Phi_2, \cdots, \Phi_K)$. Ψ is given by

$$\Psi = \sum_{k=2}^{K} Z_k \tag{2-6}$$

and the Z_k are identically distributed and conditionally independent given B, with each Z_k specified by

$$Z_k = U_k V_k \tag{2-7}$$
$$U_k = 1 + cos\,(2\Phi_k) \tag{2-8}$$
$$V_k = (2B + 1)(S_k{}^2 - S_k) + N/2 \tag{2-9}$$

The direct evaluation of (2-4) in [6] entails finding the densities of U_k and V_k conditioned on B which yields the conditional density of each Z_k, $f_{Z|B}(z)$; evaluating the $(k-2)$-fold convolution; and taking the expectation with respect to B:

$$f_\Psi(z) = E\,[f_{Z|B}(z) * \cdots * f_{Z|B}(z)] \tag{2-10}$$

Note that (2-4) is the expectation of the function $Q\,[N/\sqrt{\Psi}]$ of the random variable Ψ.[2] A considerable computational simplification was made in [2] by using the following result: Let P be a real function of θ, a random variable with mean μ and variance σ^2. Then, an accurate approximation is

$$E\,[P\,(\theta)] \approx \frac{2}{3}P\,(\mu) + \frac{1}{6}P\,(\mu + \sqrt{3}\sigma) + \frac{1}{6}P\,(\mu - \sqrt{3}\sigma) \tag{2-11}$$

Using (2-11), (2-4) can be easily evaluated by letting μ and σ be the mean and standard deviation of Ψ. In [2] it was found that

$$\mu = (K-1)E\,(Z) \tag{2-12}$$
$$\sigma^2 = (K-1)[E\,(Z^2) - E\,(Z)^2 + (K-2)cov\,(Z_j, Z_k)] \qquad (for\ any\ \ j \neq k)\,(2\text{-}13)$$

with

---

2. The randomness in Ψ is due to the randomness in the code sequences, time and phase offsets of the (K-1) interfering users, and the randomness associated with B.

$$E(Z) = \frac{N}{2} - \frac{E(B)}{3} - \frac{1}{6} = \frac{N}{3} \qquad (2\text{-}14)$$

$$E(Z^2) = \frac{1}{40}[8E(B^2) + (8-20N)E(B) + 2 - 10N + 15N^2]$$

$$= \frac{7N^2 + 2N - 2}{40} \qquad (2\text{-}15)$$

$$cov(Z_j, Z_k) = var(B)/9 = \frac{N-1}{4} \quad (for\ any\ j \neq k) \qquad (2\text{-}16)$$

$$E(B) = \frac{N-1}{2} \qquad (2\text{-}17)$$

$$E(B^2) = \frac{N(N-1)}{4} \qquad (2\text{-}18)$$

This results in

$$\hat{P}_e \approx \frac{2}{3}Q\left[\sqrt{\frac{3N}{(K-1)}}\right] + \frac{1}{6}Q\left[\frac{N}{((K-1)N/3 + \sqrt{3}\sigma)^{0.5}}\right] + \frac{1}{6}Q\left[\frac{N}{(K-1)N/3 - \sqrt{3}\sigma)^{0.5}}\right]$$

$$+ \frac{1}{6}Q\left[\frac{N}{(K-1)N/3 - \sqrt{3}\sigma)^{0.5}}\right] \qquad (2\text{-}19)$$

with

$$\sigma^2 = (K-1)\left[N^2\frac{23}{360} + N(\frac{1}{20} + \frac{K-2}{36}) - \frac{1}{20} - \frac{K-2}{36}\right] \qquad (2\text{-}20)$$

Remark. These results use certain assumptions, such as rectangular pulses and random chip sequences. Since we are interested in making relative comparisons in the sequel, the specific assumptions are less important than if we were making absolute calculations.□

3. EXTENSION TO UNEQUAL POWERS

To extend the result of the last section to the case of unequal powers, (2-4) and (2-6) are modified to

$$\hat{P}_e(p_1) = \int_0^\infty Q\left[\frac{\sqrt{p_1}\,N}{\sqrt{\Psi}}\right] f_\Psi(z)\,dz \qquad (3\text{-}1)$$

and

$$\Psi = \sum_{k=2}^{K} p_k\,Z_k \qquad (3\text{-}2)$$

where p_k is the power of the k'th user divided by 2. We shall assume here, for simplicity, that the p_k are independent, identically distributed random variables.

Remark. In the following, we shall take expectations with respect to random variables, including the p_k. An interpretation of the result would be that there would be differences in performance received by the different users and the calculation is the average of the different performances. A more benign interpretation is that, in the long run, each user receives the same average performance. That is, the powers are actually random processes (slowly varying relative to the bit rate) rather than random variables. This is more benign because the first interpretation consigns some users to permanently poor performance. The latter interpretation is appropriate here as power control will be continually adjusting the powers.□

Now, we will uncondition $\hat{P}_e$ first with respect to Ψ and then with respect to p_1:

$$\hat{P}_e = E_{P_1} E_\Psi Q\left[\frac{\sqrt{p_1}N}{\sqrt{\psi}}\right] \tag{3-3}$$

The expectations are taken using (2-11) twice:

$$\hat{P}_e = \frac{2}{3} F[E(p_1)] + \frac{1}{6} F[E(p_1)+\sqrt{3}\sigma(p_1)] + \frac{1}{6} F[E(p_1)-\sqrt{3}\sigma(p_1)] \tag{3-4}$$

with

$$F(x) = \frac{2}{3}Q\left[\frac{\sqrt{x}N}{\sqrt{E(\psi)}}\right] + \frac{1}{6}Q\left[\frac{\sqrt{x}N}{\sqrt{E(\psi)+\sqrt{3}\sigma(\psi)}}\right] + \frac{1}{6}Q\left[\frac{\sqrt{x}N}{\sqrt{E(\psi)-\sqrt{3}\sigma(\psi)}}\right] \tag{3-5}$$

and where $\sigma(\cdot)$ is the standard deviation of its argument. The mean and variance of Ψ are given by:

$$E(\Psi) = (K-1) E(p_j) E(Z) \tag{3-6}$$

$$\sigma^2(\Psi) = (K-1)\left[E(p_j)^2\sigma^2(Z)+\sigma^2(p_j)E(Z^2)+(K-2)E(p_j)^2 cov(Z_j,Z_k)\right]$$

$$(for\ any\ j\neq k) \tag{3-7}$$

The moments of Z are given in (2-14)-(2-18). The moments of p_j are parameters to exercise to see their effect on needed capacity reductions. This is the subject of the next section.

4. CAPACITY REDUCTIONS FOR THE UNCODED CASE

To exercise the results of the last section, pose the following question: If the system can support a certain number of users at a given bit error probability, how many fewer can be supported if the powers are unequal? So, suppose the powers, $p_k,\ k=1,...K$, are independent identically distributed random variables with the same mean (an optimistic assumption) and coefficient of variation (standard deviation/mean), C. C is a normalized RMS power variation. The following table gives illustrative results for $N=63$ chips/bit and an objective bit error probability of 0.001. The table was generated by first using (2-19) (or, equivalently, (3-4) with $\sigma(p_j)=0$) to see how many users, K, can be supported with equal powers. Then (3-4) was used to see how much K has to be reduced as a function of $\sigma(p_j)$ to maintain the same error probability. The last column gives the % reductions in K.

Table I

% Reduction in # Users		
RMS Pwr Variation (%)	RMS Pwr. Variation (db)	%Reduction
0	0	0
10	-0.4,+0.4	2
20	-1,+0.8	10
30	-1.5,+1.1	25
40	-2.2,+1.5	50

Remark 1. The plus and minus values of db are due to the nonlinear translation from plus and minus linear variations.□

Remark 2. The % reduction results are not too sensitive to the number of chips/bit but are sensitive to the objective error probability, increasing with objective error probability decreases. □

Remark 3. Here, each power is being controlled to a given objective power level. If one controls to the (moving) average of all the powers, this would be more difficult because of the interactions among the powers.□

5. CONVOLUTIONAL CODING

The previous results used a simple, accurate approach for the uncoded case. There are no analogous results for convolutional coding. Therefore, an upper bound on the bit error probability for soft decision Viterbi decoding will be used.

The following upper bound is given on p. 462 of [9]:

$$P_{ub}(\gamma_b) = \frac{1}{2} \sum_{d=d_{free}}^{\infty} \beta_d \, erfc(\sqrt{\gamma_b R_c d}) \qquad (5\text{-}1)$$

where

$$\gamma_b = received \ SNR \ per \ bit \qquad (5\text{-}2)$$
$$R_c = code \ rate \qquad (5\text{-}3)$$
$$\beta_d = a_d f(d) \qquad (5\text{-}4)$$

and where the a_d are the coefficients in the series expansion for the code transfer function and the $f(d)$ are the exponents of N in the code transfer function when a factor N is introduced into all the branch transitions caused by input bit 1.[3] See Chapter 5 of [9] for an exposition of this and the assumptions underlying (5-1) (assumptions not claimed to represent any real wireless system). Dependence on γ_b is explicitly shown as that will be affected by power variations.

Measurement Interval. To consider power variations, we will consider variations in the SNR. Since we are particularly interested in relating *measured* power variations to performance, we define the measurement interval as the time interval over which power measurements are averaged. The coefficients of variation needed in our calculations are obtained from the sample variance of the measurement interval averages. For the convolutional decoder under consideration, we assume the measurement interval is over the bits in the decoder memory. We then assume that the SNR is essentially constant over the decoder memory (but varying over a longer time scale). It may be shown that this will give optimistic results for this decoder when we evaluate the expression to be derived using variances obtained with the above mentioned measurement interval. If the measurement interval is longer, then the results get increasingly optimistic (unless corrected for). If the measurement interval is shorter, the use of those variations becomes pessimistic. The point to be made is not that our measurement assumption should be followed but that, in verifying power control, the assumptions on the power variations being referred to need to be specified and related to the actual system under consideration (more complex than this simple model).□

To use (5-1) to estimate the effect of unequal powers, observe that γ_b is approximately proportional to $p_1 / \sum_{k=2}^{K} p_k$. Furthermore, observe that the prime

---

3. For very low γ_b, this upper bound can exceed 1, so we actually used $\min \{1, P_{ub}(\gamma_b)\}$.

contributor to the coefficient of variation of γ_b is the variation in p_1 since the coefficient of variation of the sum in the denominator is reduced.[4] More specifically, with cv denoting coefficient of variation,

$$cv^2\left[\frac{p_1}{\sum_{k=2}^{K} p_k}\right] \approx cv^2(p_1) + cv^2(\sum_{k=2}^{K} p_k)$$

$$= cv^2(p_1)\left[1 + \frac{1}{K-1}\right] \tag{5-5}$$

Then, the following estimate is made, using (2-11):

$$\hat{P}_{ub} = \frac{2}{3}P_{ub}(E(\gamma_b)) + \frac{1}{6}P_{ub}[E(\gamma_b)(1+\sqrt{3}C)] + \frac{1}{6}P_{ub}[E(\gamma_b)(1-\sqrt{3}C)] \tag{5-6}$$

with (5-1) being used and where $E(\gamma_b)$ and C are the mean and coefficient of variation of γ_b, respectively.

Remark 1. We tested this simplified approach to estimate the effect of unequal powers against the more accurate approach of Section 4 for the uncoded case. It was found to be acceptable with some underestimation of the capacity reductions (as expected). The key observation to make is that while there is an averaging effect with the interfering users, the effect of variations in p_1 itself is most significant.□

Remark on Accuracy of (5-6). (2-11) is normally quite accurate. To verify that in this specific application (especially for a relatively large coefficient of variation of 0.4), (5-6) was also compared against the following and found to give close capacity reductions:

$$P_{ub}' = \int_0^{\infty} P_{ub}(E(\gamma_b)x) \frac{\lambda}{\Gamma(r)}(\lambda x)^{r-1}e^{-\lambda x}dx \tag{5-7}$$

with the parameters of the gamma density, λ and r, chosen to yield a mean of 1 and a coefficient of variation C. That is, a gamma density was used as a specific case for the SNR variation (gamma rather than normal to avoid negative arguments). In the use of both (5-6) and (5-7), the average SNR was increased with C to keep the error probability upper bound fixed at

---

4. This was borne out by closer inspection of the results for the uncoded case.

0.001.□

Calculations for capacity reductions were calculated analogously to those of the last section. Illustrative results for $R_c = 1/3$ and length 3 code and a 0.001 objective error probability are given in Table II.

Table II

% Reduction in # Users		
RMS Pwr Variation (%)	RMS Pwr. Variation (db)	%Reduction
0	0	0
10	-0.4,+0.4	3
20	-1,+0.8	13
30	-1.5,+1.1	32
40	-2.2,+1.5	55

Remark 1. Since the upper bound of (5-1) has a steeper P_b (probability of error) vs. *SNR* characteristic than the actual curve, we reduced the capacity reductions following from (5-6) by 2-5 percent (increasing with RMS power variation) for Table II.[5] These were taken as reasonable reductions which keep the coded case reductions more than those for the uncoded results of the last section (the actual coded curves are steeper than those for the uncoded case). It should be noted, however, that the RMS power variations (averaged over a measurement interval) in Table II are not the same as given in Table I. So, it should not be concluded that coding increases the sensitivity to power variations (the opposite is true).□

Remark 2. We emphasize again the important role of the measurement interval. If the *actual* measurement interval was smaller (or larger) than our assumption, the coefficient of variation C obtained from the measurements would be larger (or smaller) than the C in Table II.□

6. HOW WELL CAN POWER CONTROL BE DONE?

The purpose of this section is to give some insight into the problem of closed loop control. Suppose each user's power before feedback control has

---

5. Before the capacity reductions were so reduced, they were 4.5, 18, 37, and 60%.

random deviations with rms value β and the normalized autocovariance function of the power is $R(\tau)$. Suppose, further, there is a linear feedback controller which has a time lag (pure time delay) τ. This is the delay in the information on which the control actions are based; it is not the time between control actions. This time lag represents the time from when the mobile transmits to the time when a power adjustment can next be made to the mobile's transmission. Thus, the base station must receive and process the received signal and send back a power control command.

We make the simplifying assumption that the controller adjusts the reverse link power at time t based on a perfect (noiseless) measurement of the power at time $t-\tau$ seconds. We will give estimates of the ability of the feedback controller to effectively combat Rayleigh fading for mobiles (assuming one tries to do this, rather than just increasing the signal-noise requirements).

For this feedback controller, it can be shown that (e.g., [10])

$$\frac{rms\ error\ in\ controlled\ power}{rms\ deviation\ in\ uncontrolled\ power} \geq \sqrt{1-R^2(\tau)} \qquad (6\text{-}1)$$

(6-2) shows a basic limitation in how effective the feedback control can be. A few examples are given. We will use results from Chapter 1 of [11] and a maximum Doppler frequency of 40 hz (e.g., around 900 MHz, 30 mph). The first example is $\tau = 1\ msec$ which gives 0.25 for the right hand side of (6-1). Thus, if the reverse link powers can be changed based on information on those powers not older than 1 $msec$, there is potential for sizable reduction of the variations. Note that this assumes an otherwise perfect controller. That is, the controller uses otherwise perfect (noiseless) measurements and makes continuous (not quantized) corrections continuously in time (not periodically). Deviations from these assumptions raise the lower bound. If $\tau = 2\ msec$, the minimum reduction factor is 0.48. At $\tau = 4\ msec$, the minimum reduction factor is 0.89. It is seen that the potential for improvement is quite sensitive around a few milliseconds. This sensitivity is with respect to both τ for a fixed $R(\tau)$ and to changes in $R(\tau)$ for a fixed τ.

Note that the above are just examples using a number of simplifying assumptions and a particular $R(\tau)$; they are not general guidelines. In addition, the calculations can be quite sensitive. More refined studies need to be done for specific systems. What must be investigated is the combined effect of all the techniques employed to mitigate the Rayleigh fading (or whatever characterization best fits the fading).

7. CONCLUDING REMARKS

The purpose of the analyses presented is to clearly indicate certain basic sensitivities and restrictions. The models (and the specific numerical results) do not represent any complete system. Further work is needed to analyze particular configurations of more complete systems. It should be noted that particular configurations (e.g., using interleaving or diversity) can have different sensitivity to power variations. An example of a richer model is given in [12] where interleaving is taken into account (not considered in the present paper). The results of [12] indicate that the combination of power control, interleaving, and convolutional coding has the potential of providing robust protection against power variations. This paper and [12] are part of an effort to develop a sequence of models of increasing complexity, providing insights in the process. Also, consideration must be given to other effects such as measurement noise and transients in moving between base stations and in turning corners.

Another approach to power variations is multiuser detection ([13], [14]). The results of this paper assume single user detection in which other users are considered noise. An alternative being studied is to jointly detect all the users. Then all of the users become "signals" and the effects of interfering users can be partially cancelled. The issues of practical and economic feasibility need to be better understood.

Acknowledgement

I thank D. J. Goodman and A. J. Viterbi for their comments.

REFERENCES

1. K. S. Gilhousen, I. M. Jacobs, R. Padovani, A. J. Viterbi, L. A. Weaver, and C. E. Wheatley, "On the Capacity of a Cellular CDMA System," to appear in *IEEE Trans. on Vehicular Technology*, November 1990.

2. J. M. Holtzman, "A Simple, Accurate Method to Calculate Spread Spectrum Multiple Access Error Probabilities," *ICC '91*, Denver, Colorado, June 1991 (to appear in *IEEE Trans. Comm.*).

3. C. L. Weber, G. K. Smith, and B. H. Batson, "Performance Considerations of Code Division Multiple-Access Systems," *IEEE Trans. on Veh. Tech.*, Vol. VT-30, No. 1, pp. 3-10, February 1981.

4. G. L. Turin, " The Effects of Multipath and Fading on the Performance of Direct-Sequence CDMA Systems," *IEEE Journal on Selected Areas in Communications*, Vol. SAC-2, No. 4, pp.597-603, July 1984.

5. E. A. Geraniotis and M. B. Pursley, "Performance of Coherent Direct-Sequence Spread Spectrum Communications over Specular Multipath Fading Channels," *IEEE Trans. on Communications*, Vol. COM-33, No. 6, pp. 502-508, June 1985.

6. R. K. Morrow and J.S. Lehnert, "Bit-to-Bit Error Dependence in Slotted DS/SSMA Packet Systems with Random Signature Sequences," *IEEE Trans. Commun.*, vol. COM-37, no. 10, pp. 1052-1061, October, 1989.

7. J. S. Lehnert and M. B. Pursley, "Error Probabilities for Binary Direct-Sequence Spread Spectrum Communications with Random Signature Sequences," *IEEE Trans. Commun.*, vol. COM-35, no. 1, pp.87-98,

8. M. B. Pursley, "Performance Evaluation for Phase-Coded Spread-Spectrum Multiple-Access Communication - Part I: System Analysis," *IEEE Trans. Commun.*, vol. COM-25, no. 8, pp.795-799, August 1977.

9. J. G. Proakis, "Digital Communications," Second Edition, Mc-Graw-Hill Book Co., New York, 1989.

10. G. C. Newton, Jr., L. A. Gould, J. F. Kaiser, "Analytical Design of Linear Feedback Controls," John Wiley & Sons, New York, 1957.

11. W. C. Jakes, Editor, "Microwave Mobile Communications," John Wiley, New York, 1974.

12. F. Simpson and J. M. Holtzman, "CDMA Power Control, Interleaving, and Convolutional Coding," *41st IEEE Vehicular Technology Conf.*, St. Louis, May 19-22, 1991.

13. R. Lupus and S. Verdú, "Near-Far Resistance of Multiuser Detectors in Asynchronous Channels," *IEEE Trans. on Commun.*, vol. 38, no. 4, pp.496-508, April 1990.

14. M. K. Varanasi and A. Aazhang, "Multistage Detection in Asynchronous Code-Division Multiple-Access Communications," *IEEE Trans. on Commun.*, vol. 38, no. 4, pp.509-519, April 1990.

Index